Basic Mathematics

Third Edition

by James A. Smith

Book 4

Unit V
Division of
Whole Numbers

Continuing Education Systems, Inc.

1990 Edition

Basic Mathematics, Third Edition, is a major revision of *Basic Mathematics* by Daniel Bobrow. Both the original and this revision were produced under the direction of Dr. Jack E. Forbes, Professor of Mathematics, Purdue Unversity — Calumet Campus.

Published by Continuing Education Systems, Inc. 112 S. Grant St., Hinsdale, Illinois 60521

ISBN 0-916780-03-1

Contents

How to Use This Book

This is a different kind of textbook called a *programmed textbook.* It is designed to help you learn mathematics by giving you small bits of information, then giving you the review, restatement, and drill usually given in class discussions and homework.

You should note that the program is not written in paragraph form but as a sequence of numbered sentences or groups of sentences. Each numbered part is called a *frame*. An example of a frame is:

3. The title of this program is *Basic* __________________.

Mathematics

You can see that there is a "blank" in this sentence. If you read the Title Page, you know that the word "Mathematics" completes the sentence correctly. You should also notice that the word "Mathematics" is written below the heavy black dashes below the frame. This tells you how you should "fill in" the blank in the frame.

To use this program, you should:

(1) Cover the page with a sheet of paper.
(2) Move the paper down to the first set of heavy black dashes.
(3) Read the frame completely.
(4) Decide on your response—that is, the word, words, or numbers which correctly complete the sentence.
(5) Write your response on the sheet of paper (or in the book as you are instructed).
(6) Move your sheet of paper down to the next set of heavy black dashes.
(7) Check your written response with the correct response written below the heavy black dashes.
  (a) If your response was correct, go on to the next frame.
  (b) If your response was incorrect, *stop and think*! Maybe you have found just another way of saying the same thing. If not, re-read the frame and see why you made the wrong response. Then go on to the next frame.

After an idea has been taught, there will be instructions to do an exercise list. You should do the exercise list as you would do a list from a usual textbook. You should then check your answers. Answers to all exercise lists are found in the back of the book.

After you have completed from 3 to 5 exercise lists, you will find a progress test. Work the progress test and then check your answers with those provided in the back of the book. If you do not do well on the progress test, you probably should (and may be instructed to) re-do some of the program.

If you use the program properly and remember its purpose—to replace not only the textbook but much more of the usual classroom work—you will find it a quick and easy way to study *Basic Mathematics*.

Good luck.

Unit V - Division of Whole Numbers

5.1 The Meaning of Division

In this unit, we review the operation division. We begin with the definition of 7 divided by 3. We write either $3\overline{)7}$ or $7 \div 3$ to mean "7 divided by 3" or "divide 7 by 3."

To do $7 \div 3$ means that we start with a set of 7 objects.

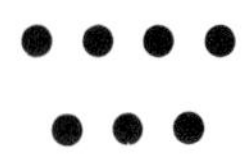

Then, to divide by 3 means that we determine the number of "sets of 3" that can be formed from the 7 objects.

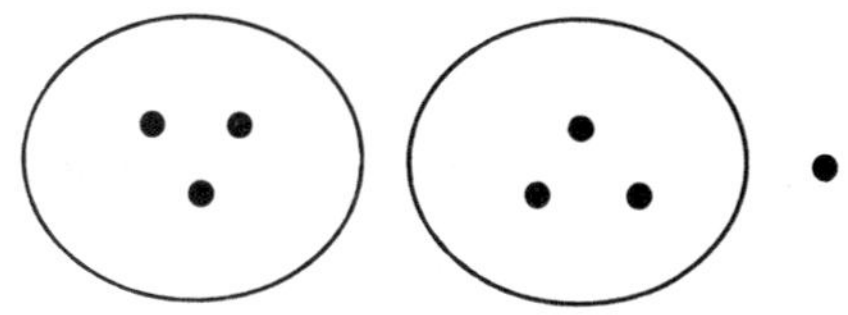

We see that we can form 2 "sets of 3" with 1 object "left over." We write $7 \div 3 = 2 \text{ R } 1$ or $\begin{array}{r} 2 \text{ R } 1 \\ 3\overline{)7\phantom{\text{ R } 1}} \end{array}$. Either $7 \div 3 = 2 \text{ R } 1$ or $\begin{array}{r} 2 \text{ R } 1 \\ 3\overline{)7\phantom{\text{ R } 1}} \end{array}$ is read "7 divided by 3 is (or equals) 2 with a remainder of 1.

1. Let's do $12 \div 5$. The 12 tells us to start with __________ objects.

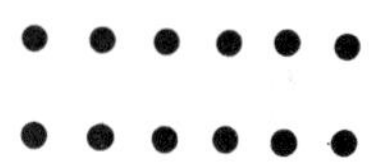

12

2. We're doing $12 \div 5$. The 5 tells us to see how many "sets of __________" we can form from the __________ objects.

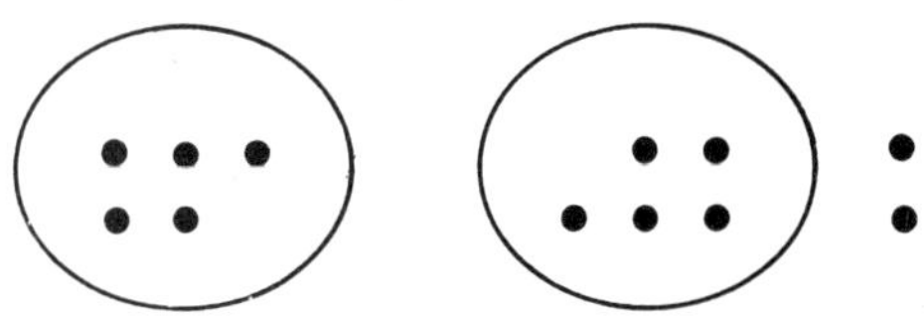

5; 12

3. Since we can form __________ "sets of 5" with __________ objects "left over,"

we write $5 \overline{)12}$ with 2 R 2 above, or 12 ÷ 5 = __________.

— — —

2; 2; 2 R 2

4. We know that $3 \overline{)7}$ tells us to divide __________ by __________.

— — —

7; 3

5. In a division problem, the number "being divided" is called the *dividend*. Thus, in $3 \overline{)7}$, __________ is the *dividend*.

— — —

7

6. In 12 ÷ 5, __________ is the *dividend*. In 14 ÷ 6, 14 is the __________.

— — —

12; dividend

7. In terms of sets of objects, the number of objects we start with is the __________.

— — —

dividend

8. In 8 ÷ 2, __________ is the dividend. In $7 \overline{)42}$, 42 is the __________.

— — —

8; dividend

9. In a division problem, the number by which we are dividing is called the *divisor*. Thus, in 7 ÷ 3, __________ is the *divisor*.

— — —

3

10. In $5\overline{)12}$, __________ is the *divisor*. In $6\overline{)14}$, 6 is the __________.

5; divisor

11. In terms of sets of objects, the number of objects in each set we form is the __________.

divisor

12. In $2\overline{)8}$, __________ is the divisor. In $42 \div 7$, 7 is the __________.

2; divisor

13. In $24 \div 6$, __________ is the dividend, and __________ is the divisor.

24; 6

14. In $32 \div 8$, 32 is the __________, and __________ is the divisor.

dividend; 8

15. In $9\overline{)45}$, __________ is the dividend, and 9 is the __________.

45; divisor

16. In $4\overline{)36}$, 36 is the __________, and 4 is the __________.

dividend; divisor

17. Now let's do 22 ÷ 4. The ____________, 22, tells us to start with ________ objects.

---

dividend; 22

18. We're doing 22 ÷ 4. The ____________, 4, tells us to form "sets of ________" from our set of 22 objects.

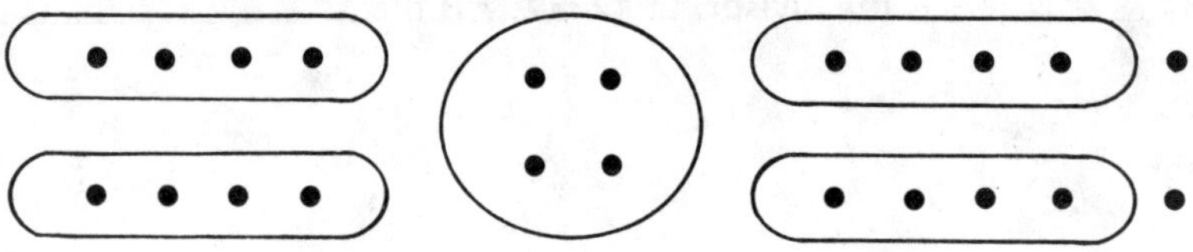

---

divisor; 4

19. Since we can form ________ "sets of 4" from our set of 22 objects with ________ objects "left over," we have 22 ÷ 4 = ________.

5; 2; 5 R 2

20. When we form "sets of 4" from our set of 22 to do 22 ÷ 4, the number of sets we form is called the *quotient*. Thus, in 22 ÷ 4 = 5 R 2, the *quotient* is ________.

5

21. In $\begin{array}{r} 2\text{ R }2 \\ 5\overline{)12} \end{array}$, the *quotient* is ________. In 7 ÷ 3 = 2 R 1, 2 is the ____________.

2; quotient

22. In 27 ÷ 4 = 6 R 3, the dividend is __________, 4 is the ________________, and 6 is the ________________.

27; divisor; quotient

23. In 34 ÷ 11 = 3 R 1, 34 is the ________________, the divisor is __________, and 3 is the ________________.

dividend; 11; quotient

24. Let's do 13 ÷ 2. The dividend tells us to start with __________ objects, and the divisor tells us to form "sets of __________."

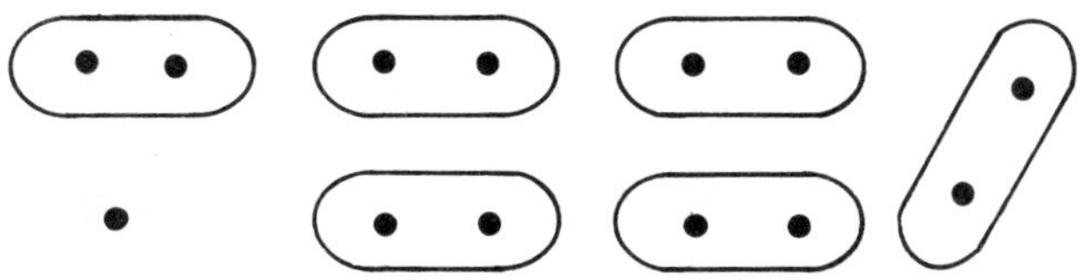

13; 2

25. Since we have formed __________ "sets of 2" and have __________ object "left over," we have 13 ÷ 2 = __________.

6; 1; 6 R 1

26. When we work a division problem using sets of objects, the dividend tells us how many objects to start with, the divisor tells us what size sets to form, and the number of sets we form is the quotient. The number of objects "left over" is called the *remainder*. Thus, in 13 ÷ 2 = 6 R 1, the *remainder* is __________.

1

27. In $\begin{array}{r} 2\text{ R }2 \\ 5\overline{)12} \end{array}$, the *remainder* is __________. In 22 ÷ 4 = 5 R 2, 2 is the __________________.

— — — — — —

2; remainder

28. In 34 ÷ 11 = 3 R 1, the quotient is __________, and 1 is the ________________.

— — — — — —

3; remainder

29. In terms of objects, a division problem looks like

$\begin{array}{r} \text{no. of sets R "no. left over"} \\ \text{no. per set}\overline{)\text{total no. of objects}} \end{array}$

In horizontal form this is
(total no. of objects) ÷ (no. per set) = ________________ R "no. left over."

— — — — — —

no. of sets

30. Using the words we have just learned, these become

$\begin{array}{r} \text{quotient R } \_\_\_\_\_\_\_\_\_\_ \\ \text{divisor}\overline{)\_\_\_\_\_\_\_\_\_\_\_\_\_\_\_\_} \end{array}$

or

(dividend) ÷ (________________) = ________________ R remainder.

— — — — — —

remainder; dividend; divisor; quotient

31. Let's do 15 ÷ 5. The dividend tells us to start with __________ objects, and the divisor tells us to form "sets of __________" from the 15 objects.

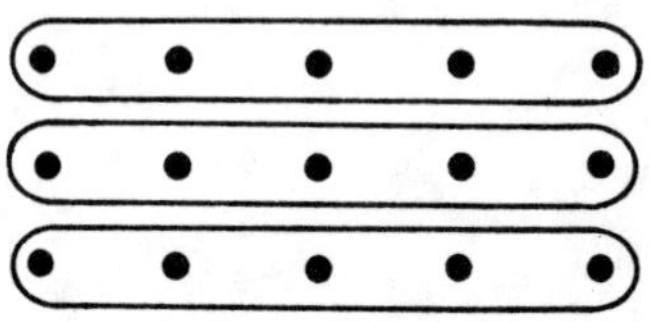

— — — — — —

15; 5

32. When we start with 15 objects and form "sets of 5," we get __________ "sets of 5" and have __________ objects "left over."

3; 0

33. Thus, when we do 15 ÷ 5, our quotient is __________, and 0 is our __________________.

3; remainder

34. We could write 15 ÷ __________ = 3 R 0. However, when 0 is our remainder, we usually don't write the R or the 0. Thus, we would write 15 ÷ 5 = __________.

5; 3

35. Use the picture below to determine 23 ÷ 4. __________

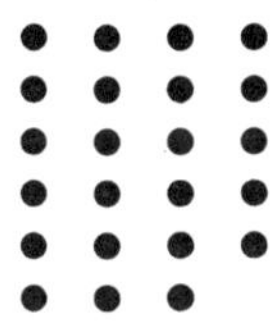

---

5 R 3

36. Use the picture below to determine 24 ÷ 6. __________

4

Now do Exercise 5.1.

5.1 Exercise List

1) Fill in the chart. (The first row is already done.)

TWO WAYS OF WRITING DIVISION		HOW THEY ARE READ
$21 \div 4$	$4\overline{)21}$	21 DIVIDED BY 4
$13 \div 5$		
	$7\overline{)31}$	
		29 DIVIDED BY 3
$37 \div 7$		

2) In $\begin{array}{r} 5\text{ R }2 \\ 7\overline{)37} \end{array}$

a) the divisor is __________,
b) the dividend is __________,
c) 5 is the ________________, and
d) 2 is the ________________.

3) In $13 \div 5 = 2$ R 3,

a) 13 is the ________________,
b) 5 is the ________________,
c) the quotient is __________, and
d) the remainder is __________.

4) Use the pictures provided to answer each division problem. Problem (a) is already begun.

a)

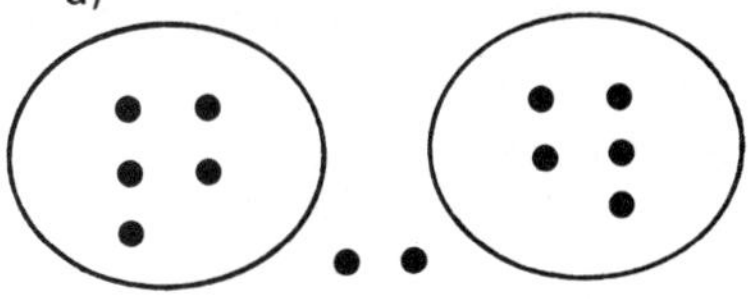

$12 \div 5 =$ ________

b)

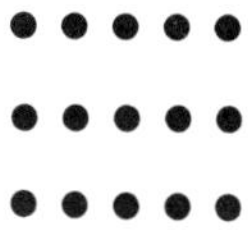

$15 \div 2 =$ ________

c)

$15 \div 3 =$ ________

d)

$7\overline{)25}$

5.2 Some Properties of Division

So far, we've seen what division means in terms of sets of objects and we've reviewed some of the words used in division problems. Of course, we don't want to draw a picture every time we work a division problem—for some problems, it would take forever just to draw the picture. So now we will look at some properties of division and see some methods for working simple division problems.

37. Below is the picture we used in frame 31.

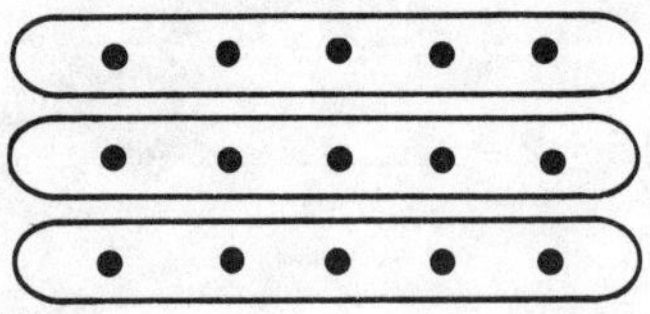

In frame 31, we were working the problem 15 ÷ 5 = ________.

--- ---

3

38. That is, we started with ________ objects and drew "ovals" to form "sets of ________."

--- ---

15; 5

39. Note that we don't need to draw "ovals." We have ________ rows with ________ dots in each row.

--- ---

3; 5

40. When we were doing multiplication, we pictured 3 "sets of 5" by drawing ________ rows with 5 dots in each row.

--- ---

3

41. Thus, to form "sets of 5" in a set of 15, we can simply draw rows of ________ until we have drawn ________ total objects.

--- ---

5; 15

42. Note that this is also a picture for 3 x 5 = __________.

— — — — — —

15

43. Below is the picture we used in frame 36.

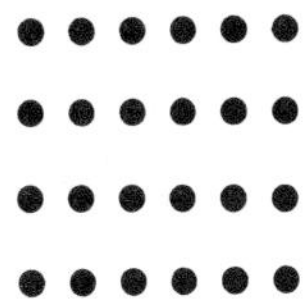

Here we have __________ objects which we have formed into "sets of __________" by drawing rows with __________ in each row. (When you did frame 36, did you draw "ovals" to form "sets of 6"? It's all right if you did, but now you should see that you didn't need to.)

— — — — — —

24; 6; 6

44. Since we used __________ rows to get a total of 24 objects, we have 24 ÷ 6 = __________.

— — — — — —

4; 4

45. Note that this is also a picture for 4 x __________ = __________.

— — — — — —

6; 24

46. Now, let's do 20 ÷ 4. We want to form "sets of __________" until we have used a total of __________ objects.

— — — — — —

4; 20

47. We can form "sets of 4" by drawing rows with ________ in each row.

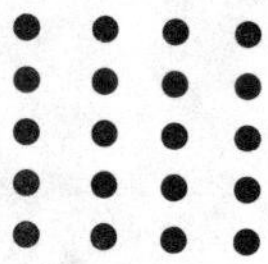

Since we need ________ rows to use 20 total objects, we have 20 ÷ 4 = ________.

— — — — — —

4; 5; 5

48. The picture in frame 47 is also a picture for ________ x 4 = ________.

— — — — — —

5; 20

49. Now suppose we consider the columns as sets in the picture of frame 47. Then we have formed "sets of 5." Since we still have 20 total objects, this is a picture of 20 ÷ 5 = ________.

— — — — — —

4

50. Consider the picture below.

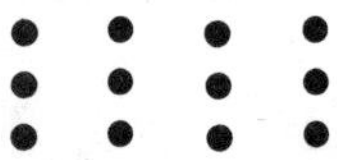

Thinking of the rows as "sets of ________," we have a picture of 12 ÷ 4 = ________. Thinking of the columns as "sets of ________," we have a picture of 12 ÷ 3 = ________.

— — — — — —

4; 3; 3; 4

51. Depending on whether we think of the rows as sets or the columns as sets, the picture in frame 50 is also a picture of 3 x 4 = ________ or 4 x ________ = ________.

— — — — — —

12; 3; 12

52. Suppose we form "sets of *b*" by drawing rows with *b* dots in each row. Suppose we have *q* rows and use a total of *a* dots. Considering the rows as "sets of *b*," we have a picture of $a \div b =$ ________ or of $q \times b =$ ________.

— — — — — —

q; *a*

53. However, if we consider the columns as "sets of *q*," we have a picture of $a \div q =$ ________ or of $b \times q =$ ________.

— — — — — —

b; *a*

54. Whenever we picture a division problem with 0 remainder in this way, we can use the same picture for another division problem and two ________ problems (in most cases).

— — — — — —

multiplication

55. Likewise, if we picture a multiplication problem in this way, we can use the same picture for another multiplication problem and two ________ problems (in most cases).

— — — — — —

division

56. Thus, we have multiplication–division "fact teams." Since 7 x 3 = ________, we get the rest of the "fact team" 3 x ________ = 21, 21 ÷ 3 = ________, and 21 ÷ 7 = ________.

— — — — — —

21; 7; 7; 3

57. From 9 x 6 = __________, we get __________ x 9 = __________, 54 ÷ 9 = __________, and __________ ÷ 6 = __________.

54; 6; 54; 6; 54; 9

58. 5 x 5 = __________. Thus, 25 ÷ 5 = __________. In this case, our "fact team" has only two members.

25; 5

59. 14 x 1 = __________. Thus, 14 ÷ 1 = __________, and 14 ÷ 14 = __________.

14; 14; 1

60. 387 x 1 = __________. Thus, 387 ÷ 1 = __________, and 387 ÷ 387 = __________.

387; 387; 1

61. In fact, if *n* is any whole number, *n* x 1 = __________.

---

n

62. This leads to two properties for division. The first of these is called the *property of 1 with respect to division.*

If *n* is any whole number, *n* ÷ 1 = *n*.

Thus, 742 ÷ 1 = __________, and 8,346 ÷ __________ = 8,346.

742; 1

63. $1\overline{)531}$ __________, $1\overline{)6{,}423}$ __________, and $1\overline{)22{,}341}$ __________.

531; 6,423; 22,341

64. The second property is called the *same number property of division*. If n is any whole number *except* 0, $n \div n = 1$.

Thus, $53\overline{)53}$ __________, and $432\overline{)432}$ __________.

1; 1

65. $983 \div$ __________ = 1, __________ $\div 3{,}216 = 1$, and $5{,}763 \div 5{,}763 =$ __________.

983; 3,216; 1

66. But wait a minute! Why did the same number property state "If n is any whole number *except* 0, $n \div n = 1$"? Shouldn't we also have $0 \div 0 = 1$? After all, we have the multiplication fact $0 \times 1 =$ __________.

0

67. However, we also have $0 \times 3 =$ __________, $0 \times 57 =$ __________, $0 \times 396 =$ __________, and, in fact, $0 \times n =$ __________ for any whole number n.

0; 0; 0; 0

68. Using the relationship between multiplication and division, this would indicate that $0 \div 0$ could be any whole number. Since there is no single number which can be $0 \div 0$, we say that $0 \div$ __________ is undefined.

0

69. Thus, there is only one division fact in the "fact team" determined by a multiplication fact with 0 as a factor. The division fact in the "team" determined by $0 \times 3 = 0$ is $0 \div 3 =$ __________.

0

70. The division fact in the "team" determined by 0 x 57 = __________ is __________________.

— — — — — —

0; $0 \div 57 = 0$ (or $57\overline{)0}$ with quotient 0)

71. For any whole number *n*, the multiplication fact $0 \times n =$ __________ yields the division fact $n\overline{)0}$ __________.

— — — — — —

0; 0

72. This is the *property of 0 with respect to division:*
If *n* is any whole number (except 0), $0 \div n =$ __________.

— — — — — —

0

73. Thus, $426\overline{)0}$ __________, $0 \div 1{,}384 =$ __________, and $23{,}469\overline{)0}$ __________

— — — — — —

0; 0; 0

74. We have already seen that $0 \div 0$ is __________________. Let's look at $7 \div 0$. Suppose we could find a whole number *a* so that $7 \div 0 = a$. Then our relationship between multiplication and division would give us __________ $\times a = 7$.

— — — — — —

undefined; 0

75. However, we know that $0 \times a =$ __________ for every whole number *a*. Since we can't have $0 \times a = 7$, $7 \div$ __________ must be undefined.

— — — — — —

0; 0

76. Suppose n is any whole number other than 0. If we had $n \div 0 = b$ for some whole number b, then we would have __________ $\times b = n$.

--- ---

0

77. Since $0 \times b =$ __________ and not n, we cannot do $n \div$ __________.

--- ---

0; 0

78. Since we already have seen that $0 \div 0$ is undefined, we say that division by 0 is *undefined*. That is, we can never divide by __________.

--- ---

0

79. Even though $0 \div 26 =$ __________, $26 \div$ __________ is undefined.

--- ---

0; 0

80. We have used multiplication facts to develop multiplication–division fact teams. From $7 \times 8 =$ __________, we also get $8 \times$ __________ $= 56$, $56 \div 8 =$ __________, and $56 \div$ __________ $= 8$.

--- ---

56; 7; 7; 7

81. We use the multiplication–division relationship to work division problems. To do

$9\overline{)36}$

we want to know how many "sets of 9" in a set of __________.

--- ---

36

82. We think, "Since

4	x	9	=	36
↑		↑		↑
no. of sets		no. per set		total

we know there are __________ 'sets of 9' in a set of 36." Thus,

$9\overline{)36}$ __________.

— — —

4; 4

83. To do
$7\overline{)63}$,
we want to know how many "sets of __________" in a set of __________.

— — —

7; 63

84. Since

__________	x	7	=	63
↑		↑		↑
no. of sets		no. per set		total

we know there are 9 "sets of __________" in a set of 63. Thus,

$7\overline{)63}$ __________.

— — —

9; 7; 9

85. To do
$8\overline{)64}$
we want to know how many "sets of __________" in a set of 64. Since

__________	x	__________	=	64
↑		↑		↑
no. of sets		no. per set		total

we know there are 8 "sets of 8" in a set of __________. Thus,

$8\overline{)64}$ __________.

— — —

8; 8; 8; 64; 8

Now do Exercise 5.2.

5.2 Exercise List

1) Fill in the chart with the rest of the members of the "fact team."

MULTIPLICATION FACTS		DIVISION FACTS	
$3 \times 7 = 21$			
	$14 \times 1 = 14$		
		$27 \div 3 = 9$	
			$36 \div 9 = 4$
$12 \times 3 = 36$			

2) a) $12 \div 12 =$ ________ b) $13 \div 1 =$ ________
c) $0 \div 4 =$ ________ d) $7 \div 0 =$ ________

3) a) $1\overline{)15}$ b) $6\overline{)0}$ c) $0\overline{)0}$ d) $7\overline{)7}$

4) Write a multiplication fact and a division fact shown by each picture.

Picture | Multiplication fact | Division fact

a) 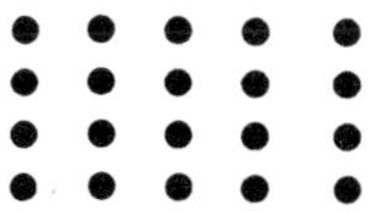 ____________________

b) 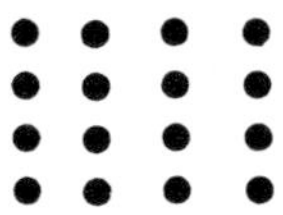 ____________________

c) 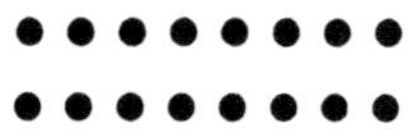 ____________________

5) Show what division must be done to solve the problem:

a) A large family with 27 people wanted to ride the ferris wheel at the carnival. If 3 people ride in each seat, how many seats will be used? ____________

b) The family in part a) wants to ride the ghost train. If 9 people fit in the train, how many trainloads will they need? ____________

5.3 Non-Zero Remainders

Now that we've seen the relationship between division and multiplication when we have 0 as remainder, let's see what happens when 0 is not the remainder.

86. Let's do 14 ÷ 4. The dividend tells us to start with __________ objects, and the divisor tells us to form "sets of __________."

14; 4

87. We can picture 14 objects formed into "sets of 4" by drawing "rows of 4" until we use 14 objects.

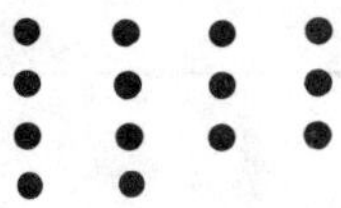

Of course, the last row contains only __________ objects.

2

88. Since we have formed __________ "sets of 4" with __________ objects "left over," we see that 14 ÷ 4 = __________.

3; 2; 3 R 2

89. Let's look again at the picture for 14 ÷ 4 = 3 R 2.

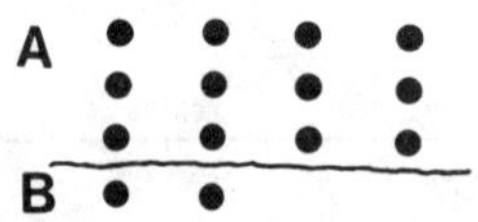

Here we have "split" the 14 objects into two sets. In A, we have 3 "rows of __________" and, in B, we have __________ dots.

4; 2

90. The number of objects in 3 "rows of 4" is 3 x __________.

4

91. Thus, we have 14 = (3 x 4) + __________.

2

92. In 14 ÷ 4 = 3 R 2, the dividend is __________, the divisor is __________, the quotient is __________, and the remainder is __________.

14; 4; 3; 2

93. Thus, for this problem, we can write
dividend = (quotient x ________________) + remainder.

---

divisor

94. Now, let's do 23 ÷ 5. The dividend tells us to start with __________ objects, and the divisor tells us to form "sets of __________."

23; 5

95. Again, we can picture 23 objects formed into "sets of 5" by drawing "rows of __________" until we use 23 objects.

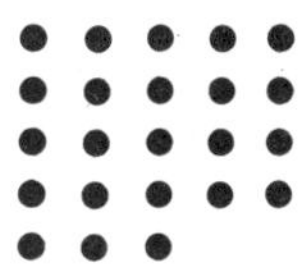

Of course, we only have __________ to put in the last row.

5; 3

96. Since we have formed __________ "sets of 5" with __________ objects left over, we see that 23 ÷ 5 = __________.

— — — — — —

4; 3; 4 R 3

97. If we look at the picture in frame 95, we have 4 "rows of __________" and 1 row containing __________ objects.

— — — — — —

5; 3

98. The number of objects in 4 "rows of 5" is __________ x 5.

— — — — — —

4

99. We see that we have 23 = (4 x __________) + 3.

— — — — — —

5

100. In 23 ÷ 5 = 4 R 3, the dividend is __________, the divisor is __________, the quotient is __________, and the remainder is __________.

— — — — — —

23; 5; 4; 3

101. Again, we see that we have
dividend = (__________ x divisor) + remainder.

— — — — — —

quotient

102. Let's see that this relationship always holds. Suppose *a* and *b* are any whole numbers and we want to do *a* ÷ *b*. (Of course, *b* cannot be __________ since we cannot divide by __________.)

— — — — — —

0; 0

103. To do $a \div b$, we start with __________ objects and form as many "sets of __________" as we can.

— — — — — —

$a; b$

104. Suppose we form q "sets of b" and have r "left over." Then we have $a \div b =$ __________.

— — — — — —

q R r

105. However, in q "sets of b," we have $q \times$ __________ objects. Since there were r objects "left over," we have a total of $(q \times b) +$ __________ objects.

— — — — — —

$b; r$

106. Since we started with a objects, we have
$a = ($__________ $\times b) +$ __________.

— — — — — —

$q; r$

107. In $a \div b = q$ R r, the dividend is __________, the divisor is __________, q is the __________________, and the remainder is __________.

— — — — — —

a; b; quotient; r

108. Thus, we again have
__________________ = (quotient x __________________) + remainder.

— — — — — —

dividend; divisor

109. Now recall the relationship between addition and subtraction. Since 17 = 9 + 8, 17 – 9 = __________.

— — — — — —

8

110. Since 13 = 7 + 6, __________ − 7 = 6.

13

111. Since, in a division problem, we have
dividend = (quotient x ______________) + ______________, this relationship between addition and subtraction gives us
dividend − (quotient x divisor) = ______________.

---

divisor; remainder; remainder

112. Thus, to work a division problem, we determine the quotient, multiply the quotient times the ______________, and subtract this number from the ______________ to determine the remainder.

---

divisor; dividend

113. So the problem is to determine the quotient! Suppose we want to do $6\overline{)34}$. Since we know that 6 is the divisor, we will multiply the quotient times __________.

6

114. Since we will multiply the quotient times 6, we think of the numbers we get when we multiply times 6. These are 0 x 6 = __________, 1 x 6 = __________, 2 x 6 = __________, 3 x 6 = __________, 4 x 6 = __________, 5 x 6 = __________, 6 x 6 = __________, and we stop here. (Recall that these numbers are called *multiples* of 6.)

0; 6; 12; 18; 24; 30; 36

115. We note that

5	x	6	=	30
↑		↑		↑
no. of sets		no. per set		total

and, since 30 is (more/less) _______________ than 34, we (can/cannot) _______________ form at least 5 "sets of 6" from a set of 34.

--- ---

less; can

116. However,

6	x	6	=	36
↑		↑		↑
no. of sets		no. per set		total

and, since 36 is (more/less) _______________ than 34, we (can/cannot) _______________ form 6 "sets of 6" from a set of 34.

--- ---

more; cannot

117. The largest number of "sets of 6" we can form from a set of 34 is __________. To organize our work, we write

$$6\overline{)34}^{\;5}$$

--- ---

5

118. Although we have already seen from the pictures that __________ is the remainder when we do 34 ÷ 6, we don't want to draw pictures every time we do a division problem. We remember that in a division problem we have

dividend – (quotient x _______________) = remainder.

--- ---

4; divisor

119. Thus, we next do 5 x 6 = ________. Since we will subtract 30 from ________, we write

$$\begin{array}{r} 5 \\ 6\,\overline{)\,34} \\ \underline{30} \end{array}$$

30; 34

120. Our work is organized to do 34 – 30. Complete the subtraction below.

$$\begin{array}{r} 5 \\ 6\,\overline{)\,34} \\ \underline{30} \\ \_\_\_\_ \end{array}$$

4

We can put together all the work.

$$\begin{array}{r} ? \\ 6\,\overline{)\,34} \end{array}$$

5 x 6 = 30 so

$$\begin{array}{rl} 5 & \leftarrow \text{sets of 6} \\ 6\,\overline{)\,34} & \\ \underline{30} & \\ 4 & \leftarrow \text{left over} \end{array}$$

121. The standard form for expressing the answer in a division problem is to now write R 4 to the right of the 5 to have

$$\begin{array}{r} 5\text{ R }4 \\ 6\,\overline{)\,34\quad} \\ \underline{30}\quad \\ 4\quad \end{array}$$

We see that 34 ÷ 6 = ________.

---

5 R 4

122. When we use sets to think about division, the ________________ tells us how many sets we form, the ________________ tells how many in each set, and the ________________ tells how many objects "left over."

--- ---

quotient; divisor; remainder

123. Suppose the remainder were greater than (or equal to) the divisor. This would mean that we could form at least one more set the size of the divisor. Thus, the remainder must always be less than the ________________.

--- ---

divisor

124. Sometimes, in division problems, we "guess" the quotient. We then multiply the "guess" times the divisor and subtract this product from the ________________. If the difference is greater than or equal to the divisor, we made a bad "guess."

--- ---

dividend

125. That is, to do 45 ÷ 7, we don't have to think 0 x 7 = __________, 1 x 7 = __________, 2 x 7 = __________, 3 x 7 = __________, 4 x 7 = __________, 5 x 7 = __________, 6 x 7 = __________, 7 x 7 = __________, and stop here.

--- ---

0; 7; 14; 21; 28; 35; 42; 49

126. We can observe that 45 is "fairly large." We might start with __________ x 7 = 35, __________ x 7 = 42, __________ x 7 = 49.

--- ---

5; 6; 7

127. $7\overline{)45}^{\,?}$

Since 6 x 7 = 42 but 7 x 7 = 49, we have

$7\overline{)45}$ ________ ← sets of 7

________ ← total used up

________ ← left over

— — —

6; 42; 3

128. Thus, 45 ÷ 7 = __________.

— — —

6 R 3

129. To do 42 ÷ 8, we might think "__________ x 8 = 32, 5 x 8 = __________, 6 x 8 = __________," and stop here.

— — —

4; 40; 48

130. $8\overline{)42}^{\,?}$

Since 5 x 8 = 40 but 6 x 8 = 48

$8\overline{)42}$ ________ ← sets of 8

________ ← total used up

________ ← left over

— — —

5; 40; 2

131. Thus, 42 ÷ 8 = __________.

— — —

5 R 2

132. To do 67 ÷ 9, we might think "6 x 9 = __________, 7 x 9 = __________, 8 x 9 = __________," and stop here. (As we get better at this, we go directly to __________ x 9 = 63.)

— — —

54; 63; 72; 7

133.

$$9\overline{)67}^{\ ?}$$

Since 7 x 9 = 63 while 8 x 9 = 72

__________ ← sets of 9
9) 67
__________ ← total used up
__________ ← left over

— — —

7; 63; 4

134. Thus, 67 ÷ 9 = __________.

— — —

7 R 4

135.

$$8\overline{)71}^{\ ?}$$

Since 8 x 8 = __________ and 9 x 8 = __________

__________ ← sets of 8
8) 71
__________ ← total used up
__________ ← left over

— — —

64; 72; 8; 64; 7

136. Thus, 71 ÷ 8 = __________.

— — —

8 R 7

137.

$$\begin{array}{r} \underline{\qquad\quad} \leftarrow \text{sets of 6} \\ 6\overline{)59} \\ \underline{\qquad\quad} \leftarrow \text{total used up} \\ \underline{\qquad\quad} \leftarrow \text{left over} \end{array}$$

9; 54; 5

138. 59 ÷ 6 = __________.

---

9 R 5

139.

$$\begin{array}{r} \underline{\qquad\quad} \\ 3\overline{)28} \\ \underline{\qquad\quad} \\ \underline{\qquad\quad} \end{array}$$

9; 27; 1

140. We then write R 1 to the right of the 9 to have

$$\begin{array}{r} 9\,\underline{\qquad\quad} \\ 3\overline{)28} \\ \underline{27} \\ 1 \end{array}$$

R 1

141.

$$\begin{array}{r} \underline{\qquad\quad} \\ 8\overline{)59} \\ \underline{\qquad\quad} \\ \underline{\qquad\quad} \end{array}$$

$$\begin{array}{r} 7\text{ R }3 \\ 8\overline{)59} \\ \underline{56} \\ 3 \end{array}$$

Now do Exercise 5.3.

5.3 Exercise List

1) a) $30 \div 7 =$ ______ b) $27 \div 4 =$ ______ c) $32 \div 6 =$ ______

2) a) $4\overline{)17}$ b) $8\overline{)35}$ c) $9\overline{)50}$ d) $3\overline{)29}$

3) a) $8\overline{)50}$ b) $8\overline{)60}$ c) $8\overline{)70}$ d) $8\overline{)75}$

4) a) $4\overline{)6}$ b) $6\overline{)4}$ c) $8 \div 3 =$ ______

d) $3 \div 8 =$ ______

5) a) $27 \div (9 \div 3)$ b) $(27 \div 9) \div 3$

6) Fill in the missing parts for each division:

	Dividend	=	Quotient	x	Divisor	+ Remainder
a) $9\overline{)56}$	56	= (	______	x	9)	+ ______
b) $4\overline{)3}$	3	= (	______	x	4)	+ ______
c) $3\overline{)}$ with quotient 7 R 1	______	= (	7	x	______)	+ 1

5.4 Division: Dividends Less Than 1,000; Divisors Less Than 10

So far in this unit, we have reviewed the meaning of division and several properties of division. We have also reviewed how we use basic multiplication facts to solve any division problem when both the divisor and quotient are less than 10.

When the dividends and quotients are larger the method we've been using gets hard—and messy.

142. Consider

$$3\,\overline{)\,45}^{\;?}$$

We want to know how many sets of ________ we can form from the set of 45. We can replace each ten by ________ ones to have 45 ones—then form sets of ________.

3; 10; 3

143.

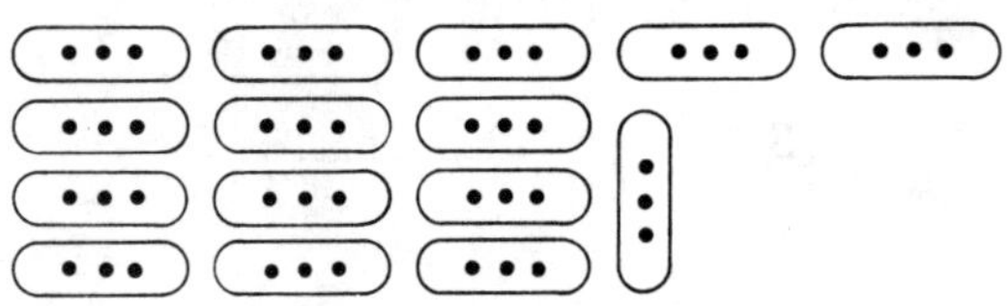

Counting, we see that there are ________ sets of 3 in a set of 45—with no objects left over. Therefore,

$3\,\overline{)\,45}$ ________.

15; 15

As you can see, this gets messy—and think what would have happened if we had 394 instead of 45! You'd be forming sets of 3 all day! But there is a much easier way to think about such problems. To see this, let's go back to some smaller numbers.

144.

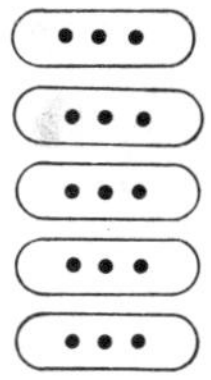

Here we have __________ sets of 3.

5

145. Now let's add some lines to the picture to picture the columns as 3 equal sets.

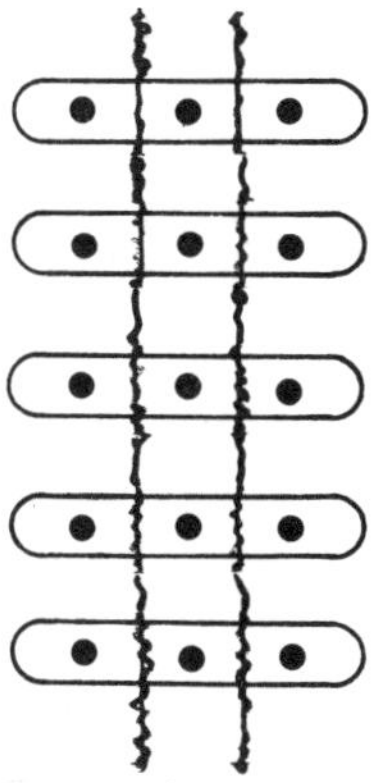

← 1 for each of 3 equal sets

← 2 for each of 3 equal sets

← 3 for each of 3 equal sets

← 4 for each of 3 equal sets

← 5 for each of 3 equal sets

We see that in each of 3 equal sets we have __________ objects.

5

146. We also see that each set of 3 gives 1 for *each* of __________ equal sets.

3

147. Since we have 5 sets of 3, we get __________ for each of 3 equal sets.

5

148. That is, the number of sets of 3 is the same as the number in each of __________ equal sets.

3

149. Suppose we had sets of 8. Each set of 8 would give 1 for each of ________ equal sets.

— — —

8

150. Thus, if we had 12 sets of 8, we would get ________ in each of 8 equal sets.

— — —

12

151. If we had 76 sets of 35, we would get 76 in each of ________ equal sets.

— — —

35

152. Earlier, we defined $15 \div 3$ as: start with ________ objects and form "sets of ________." Then the *number of sets* is $15 \div$ ________.

— — —

15; 3; 3

153. Now we see that we can also think of $15 \div 3$ as start with ________ objects and form ________ sets of "equal" size. Then the *number of objects* in each set is ________ $\div 3$.

— — —

15; 3; 15

154. If we want to do $27 \div 9$, we can think of this as start with ________ objects. We can then form "sets of ________," in which case the number of sets is $27 \div 9$, or we can form ________ sets of "equal" size. In this case the number of objects in each set is $27 \div$ ________.

— — —

27; 9; 9; 9

155. It is very handy to have both ways of thinking about division. For example, a pizza is cut into 12 pieces and is to be split equally among 3 people. How many pieces does each get? To solve this we do 12 ÷ __________ by forming __________ sets of "equal" size.

— — — — — —

3; 3

156. A grocer has 12 apples which he wants to wrap in packages of 3 apples each. How many packages will he need? Again, we solve by doing 12 ÷ 3 but here we are forming "sets of __________."

— — — — — —

3

157. Now, let's return to 45 ÷ 3. If we consider multiplying numbers times 3, we might think 7 x 3 = __________, 8 x 3 = __________, 9 x 3 = __________.

— — — — — —

21; 24; 27

158. Since 27 is less than 45, we see that there are more than 9 sets of 3 in __________.

— — — — — —

45

159. Let's draw a picture. 45 is __________ tens and __________ ones. We have

$3\overline{)45}$ 

— — — — — —

4; 5

160. Since the number of sets of 3 (quotient) is the same as the number in __________ "equal" sets, we will form __________ equal sets. We have

3) 45

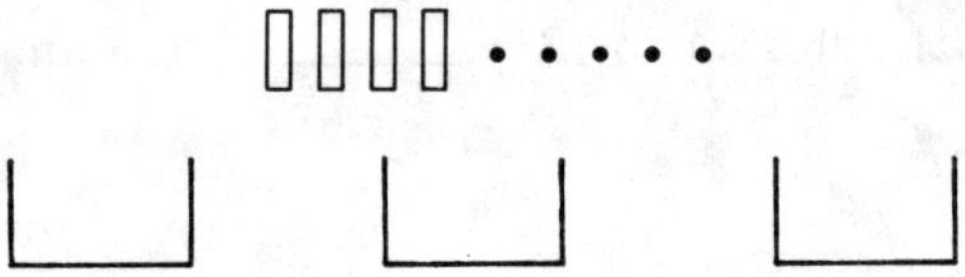

3; 3

161. Look at tens. Can we put 1 ten in each of 3 equal sets? __________________ Can we put 2 tens in each of 3 equal sets? __________________ We have

3) 45

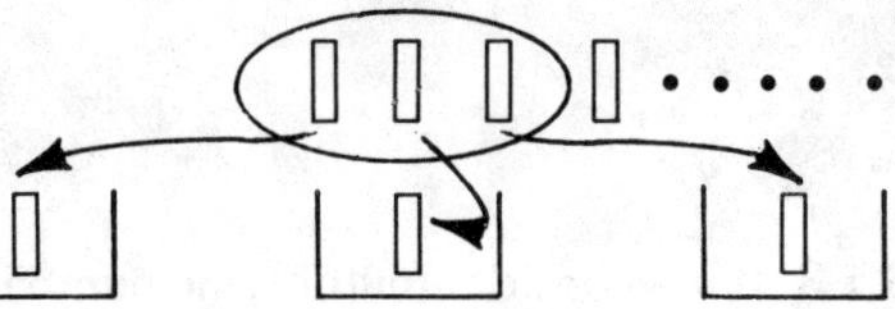

Yes; No

162. To indicate that we have 1 ten in each of our "equal" sets, we write 1 above the 4 of 45. The 4 of 45 means 4 __________________. We have

$$3\overline{)45}^{\;1}$$

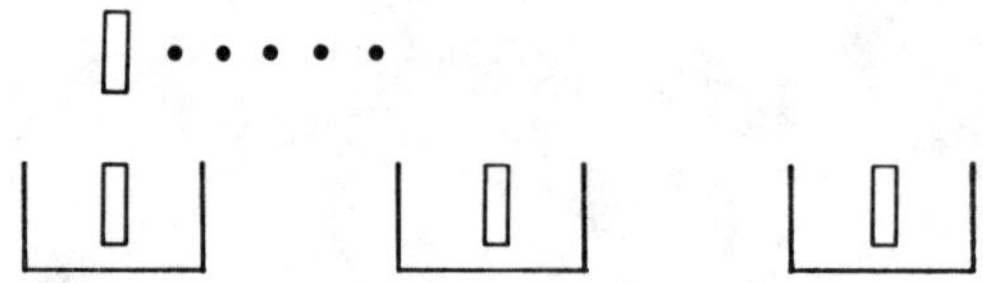

1 is called the tens *partial quotient* for the problem.

---

tens

163. Since we put 1 ten in each of our 3 "equal" sets, we have "used up" 1 x 3 = ______ of our original 4 tens. We do the multiplication 1 x ______ = 3 and write 3 below the 4 of 45 to have

```
   1
3 ) 45
    3
    -
```

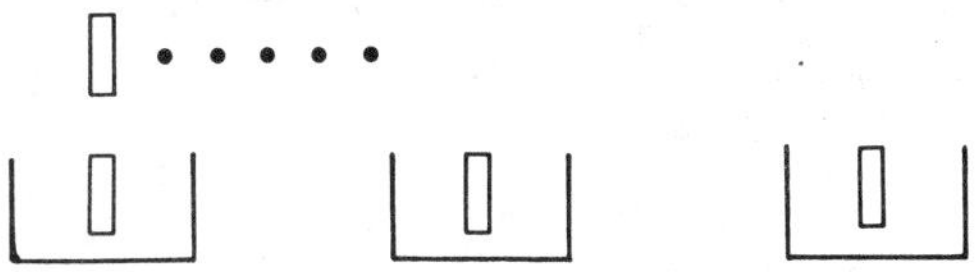

That is, we do the multiplication (tens partial quotient x ______).

3; 3; divisor

164. Since we have used 3 tens, we have (4 – ______) tens = ______ ten left. We do the subtraction within the problem to have

```
3 ) 45
    3
    -
    1
```

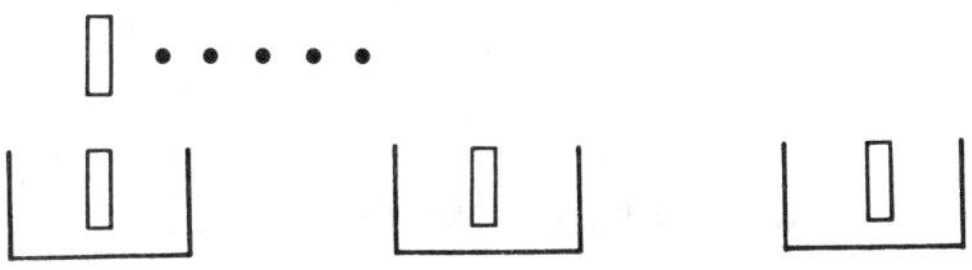

We have done tens part of dividend – (tens partial ______ x divisor).

3; 1; quotient

165. Note how we record our work.

Dividing tens:

```
                        1     ← 1 ten per set
                    3 ) 45
                        ↖
3 tens used up    →     3     4 tens started with
                        -
1 ten left over   →     1
```

The column reading 1, 4, 3, 1 from top to bottom is often called the *tens column for the problem.* Every digit in this column represents ______.

---

tens

166. We don't have enough tens left to put more in each "square" set. However, 1 ten is __________ ones. Thus, we "rename" 1 ten as 10 __________________. We indicate this within the problem by "bringing down" the 5 in the ones column to show that we are working with __________ ones. We have

$$\begin{array}{r} 1 \\ 3\,\overline{)45} \\ \underline{3} \\ 15 \end{array}$$

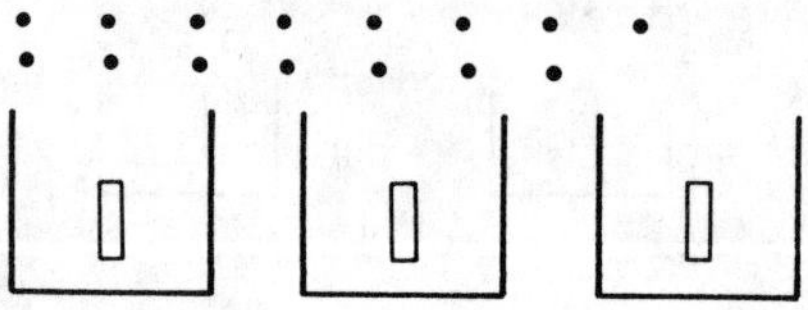

10; ones; 15

167. We have 15 ones. Since 5 x 3 = __________, we can form __________ sets of 3.

15; 5

168. However, since we can form 5 sets of 3, we can put __________ in each of 3 "equal" sets. We have

$$\begin{array}{r} 1 \\ 3\,\overline{)45} \\ \underline{3} \\ 15 \end{array}$$

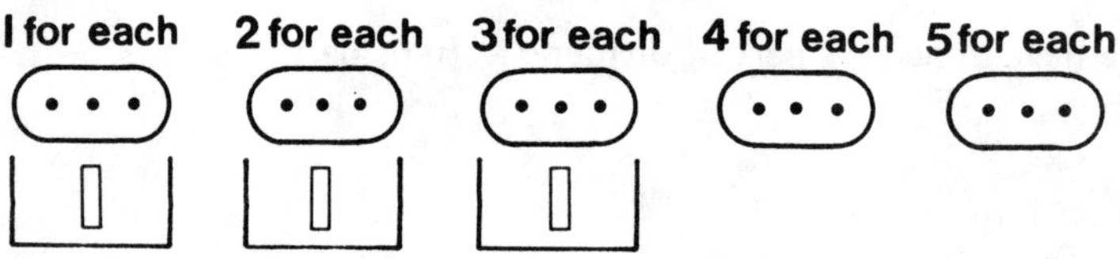

Putting them in the equal sets, we get

$$\begin{array}{r} 1 \\ 3\,\overline{)45} \\ \underline{3} \\ 15 \end{array}$$

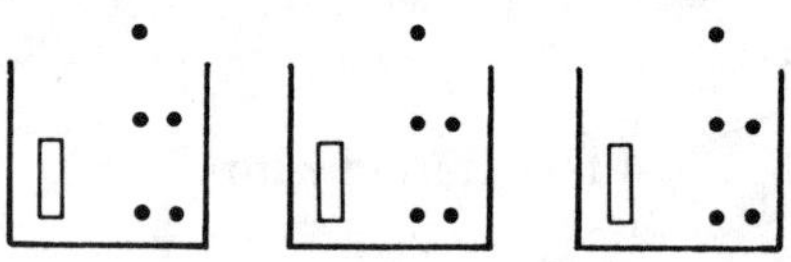

5

169. Since we put 5 ones in each of our 3 "equal" sets, we write __________ in the __________________ place of the quotient to have

```
    15
3 ) 45
    3
    15
```

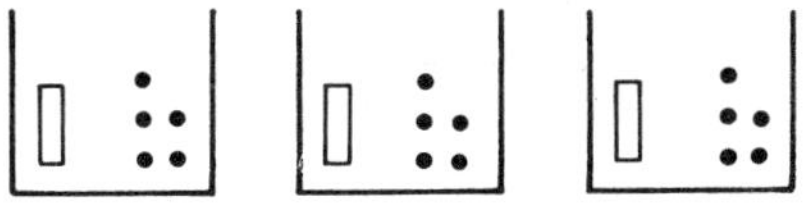

5; ones

170. Since we put 5 ones in each of our "equal" sets, we used 5 x __________ ones. We do the multiplication 5 x 3 = __________ and write 15 within the problem to have

```
    15
3 ) 45
    3
    15
    15
```

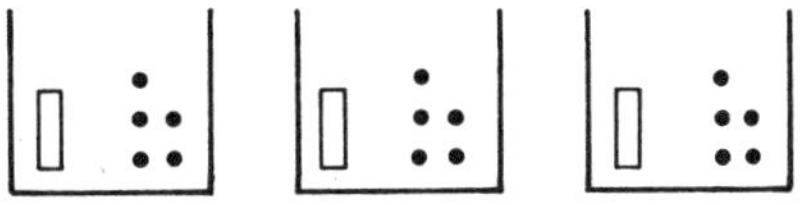

3; 15

171. Since we used 15 ones, we have (15 – __________) ones = __________ ones left. We do the subtraction within the problem to have

```
   15
3 ) 45
   3
   15
   15
    0
```

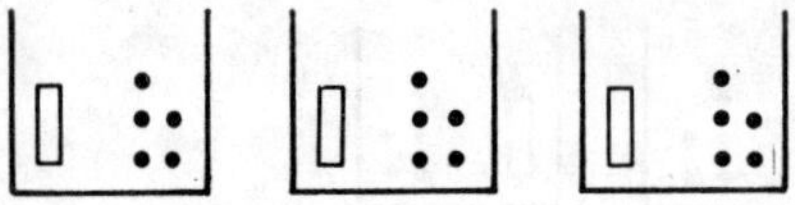

We see that 45 ÷ 3 = __________.

— — —

15; 0; 15 (or 15 R 0) Note again how we record our work.

```
                          15      5 ones per set
                       3 ) 45
                           3
15 ones ----------------> 15
15 ones used up --------> 15
0 ones left over -------> 0
```

172. Now, let's do 452 ÷ 3 without a picture. 452 tells us we have 4 ______________, 5 tens, and __________ ones. The divisor tells us to form __________ "equal" sets.

— — —

hundreds; 2; 3

173. Do we have enough hundreds to put at least one in each of 3 "equal" sets? ______________ If so, how many in each set? __________ How many hundreds have we used? __________ How many hundreds do we have left? __________

```
   ___
3 ) 452
   ___

   ___
```

— — —

```
                                      1
Yes; 1; 3(= 1 x 3); 1(= 4 – 3);  3 ) 452
                                     3
                                     1
```

174. Now compare

$$3\overline{)4} \quad \text{quotient } 1;\ 3;\ 1 \qquad 3\overline{)45} \quad \text{quotient } 1;\ 3;\ 1 \qquad 3\overline{)452} \quad \text{quotient } 1;\ 3;\ 1$$

```
   1          1           1
3 )4       3 )45       3 )452
   3          3           3
   1          1           1
```

In 4 ÷ 3, we have 4 ones; in 45 ÷ 3, we have 4 ________________; and in 452 ÷ 3, we have __________ hundreds. However, the work we do is the same. We just have more work to do in the latter two problems.

---

tens; 4

175. We can put no more hundreds in our equal sets. However, 1 hundred is __________ tens. Thus, we "rename" 1 hundred as 10 __________ and indicate this within the problem by "bringing down" the 5 to show that we now have __________ tens. We have

```
   1
3 )452
   3
   15
```

10; tens; 15

176. We have 15 tens. How many can we put in each of 3 "equal" sets? ________ We have

```
   15
3 )452
   3
   15
```

5

177. We next do 5 tens x 3 = __________ tens and subtract from the tens part of the dividend to have

```
   15
3 ) 452
    3
    15
    15
    ___
```

— — — — — —

15; 0

178. Now, comparing 45 ÷ 3 with 452 ÷ 3, we have, after two steps

```
   15          15
3 ) 45      3 ) 452
    3           3
    15          15
    15          15
     0           0
```

The first two columns within the problems are the same. However, in 45 ÷ 3, we have 15 ________________ in the second step, but in 452 ÷ 3, we have 15 ________________ in the second step.

— — — — — —

ones; tens

179. We have 0 tens left after doing 15 tens − 15 tens = __________. Thus, we don't have to "rename" any tens as ones. However, *we still have to take care of the ones in the dividend.* We "bring down" the __________ from the units place of the dividend to have

```
     15
  3 )452
     3
     15
     15
      02
```

Note: Even though 02 is written within the problem, it means we have __________ ones.

---

0 (or 0 tens); 2; 2

180. Do we have enough ones to put at least one in each of 3 "equal" sets? ________________ Thus, our ones partial quotient is __________.

We have

```
     150
  3 )452
     3
     15
     15
      02
```

No; 0

181. Now, we could do the multiplication 0 x 3 = __________ and subtract 0 from __________ to have

```
   150
3 )452
   3
   15
   15
    02
     0
     2
```

However, when a partial quotient is 0, we usually don't do the multiplication and subtraction but go on to the next step. In this case, the next step is to list the remainder to the right of the numeral for the quotient to have

```
   150 R 2
3 )452
   3
   15
   15
    02
```

We see that $452 \div 3 =$ __________.

0; 2 (or 02); 150 R 2

182. Note: A lot of people forget to write the 0 in the ones place for the quotient. However, the 0 is necessary so that the 1 means __________ hundred and the 5 means 5 __________________.

1; tens

183. Now, let's consider the problem 138 ÷ 4. The dividend tells us we have 1 __________, 3 __________, and __________ ones. The divisor tells us to form __________ "equal" sets. We have

$4\overline{)138}$

---

hundred; tens; 8; 4

184. The 1 in 138 tells us we have __________ hundred. Do we have enough hundreds to put at least one in each of 4 sets? __________ Thus, our hundreds partial quotient is __________.

1; No; 0

185. Now, we could write 0 above the 1 in 138, do our usual multiplication, and then subtract to have

$$\begin{array}{r} 0 \\ 4\overline{)138} \\ \underline{0} \\ 1 \end{array}$$

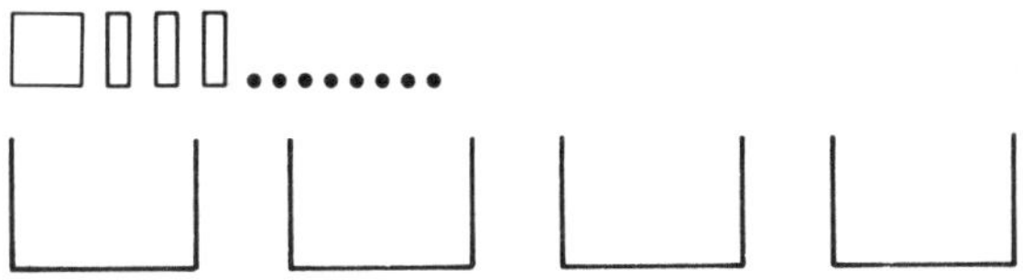

However, we know that __________ hundred is 10 __________.

1; tens

186. Also, 1 hundred and 3 tens is __________ tens. That is, the 13 of 138 can mean 13 __________.

13; tens

187. Thus, we "rename" the 1 hundred as __________ tens and look at the first *two* digits of 138 to determine our first partial quotient. We have

4) 138

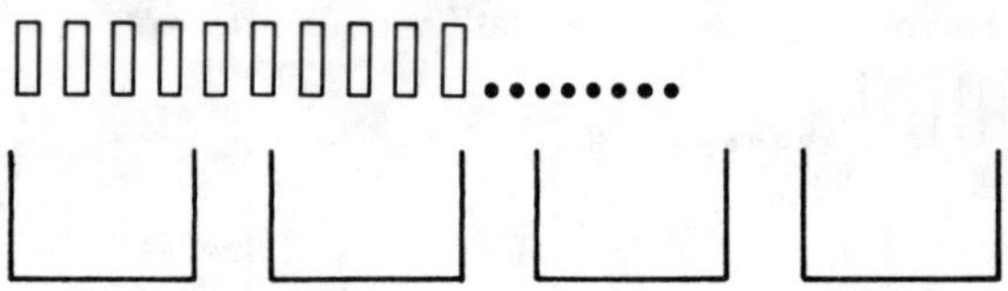

— — —

10

188. We have 13 tens. We think, "what is the largest number which we can multiply by 4 and still be less than or equal to 13?" Since 4 x 3 = __________ and 4 x 4 = __________, our tens partial quotient is __________. That is, we can put __________ tens in each of __________ "equal" sets (the "boxes" in the picture). We have

```
   3
4 ) 138
```

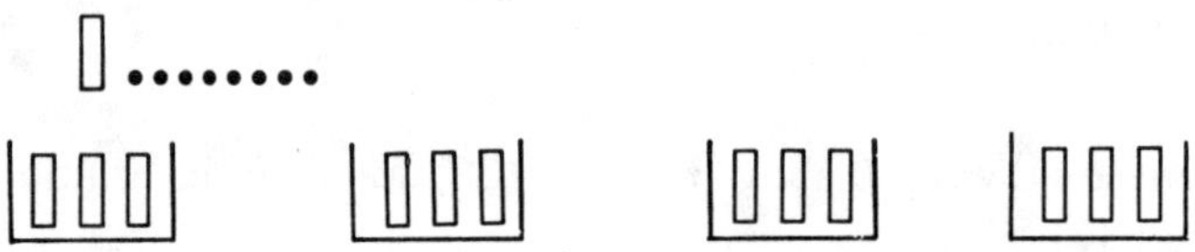

Note that the 3 for the partial quotient is written so it means 3 tens.

— — —

12; 16; 3; 3; 4

189. Although we can see from our picture that we have 1 ten remaining, we again do 3 x 4 = __________ and subtract this from __________ to have

```
   3
4 ) 138
    12
     1
```

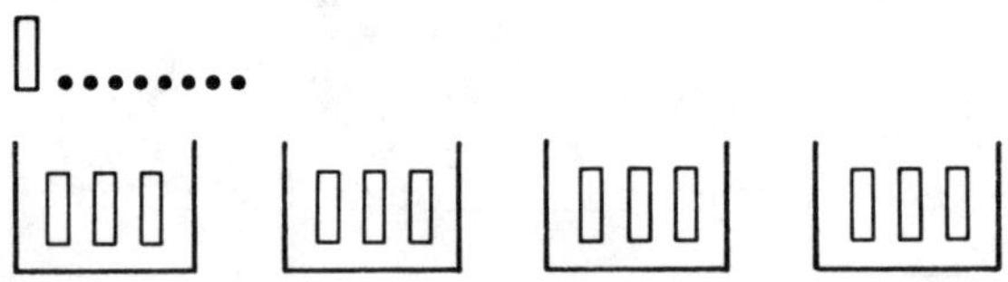

Note that we "think" 3 x 4 = 12 and 13 − 12 = 1 although we are actually doing 3 tens x 4 = __________ tens and 13 ________________ − 12 ________________ = __________ ten.

— — —

12; 13; 12; tens; tens; 1

190. We next "rename" the 1 ten as __________ ones. Within the problem, we "bring down" the 8 to show that we have a total of __________ ones. We have

$$\begin{array}{r} 3 \\ 4\,\overline{)\,138} \\ \underline{12} \\ 18 \end{array}$$

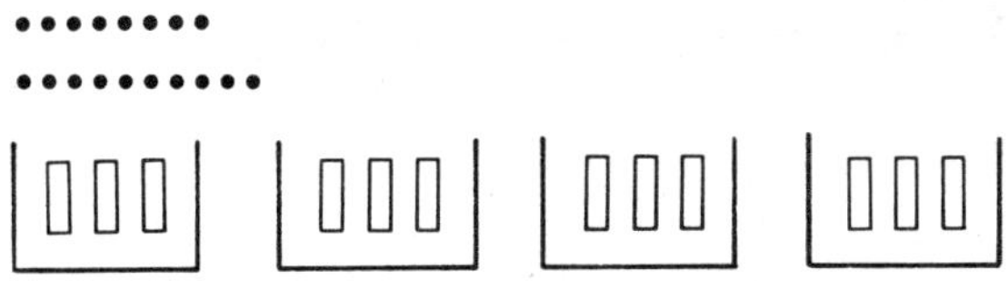

10; 18

191. We next want to see how many ones we can put in each "box." Since 4 x 4 = __________ and 4 x 5 = __________, we see that we can put __________ ones in each "box." That is, our ones partial quotient is __________. We have

$$\begin{array}{r} 34 \\ 4\,\overline{)\,138} \\ \underline{12} \\ 18 \end{array}$$

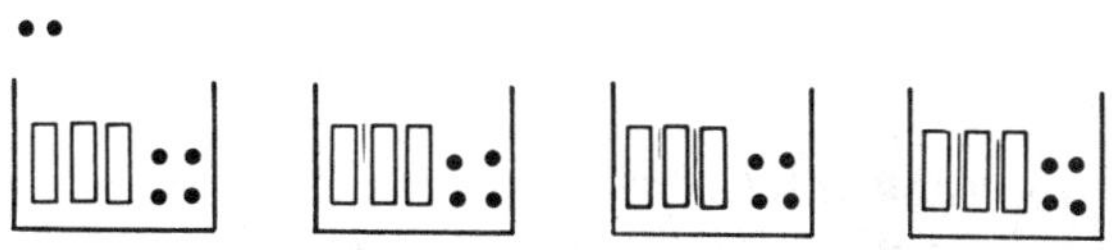

16; 20; 4; 4

192. We can see from our picture that our quotient is __________ and the remainder is __________. However, to complete the written work, we do 4 x __________ = __________ and subtract from 18 to get __________. We have

$$\begin{array}{r} 34 \\ 4\,\overline{)\,138} \\ \underline{12} \\ 18 \\ \underline{16} \\ 2 \end{array}$$

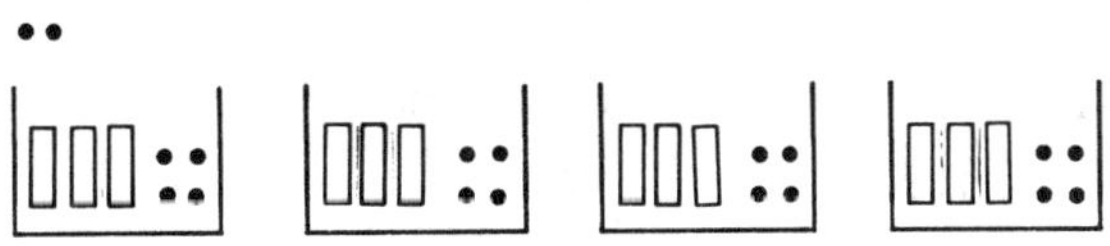

We see that 138 ÷ 4 = __________.

34; 2; 4; 16; 2; 34 R 2

193. Now let's work 753 ÷ 7 without pictures. We first write

$7\overline{)753}$

In 753, we have 7 ________________.

– – –

hundreds

194. Do we have enough hundreds to put at least one in each of 7 sets? ________________ If so, how many per set? __________ What is our hundreds partial quotient? __________ We have

$$\begin{array}{r} 1 \\ 7\overline{)753} \end{array}$$

– – –

Yes; 1; 1

195. We next multiply our hundreds partial quotient times the divisor and subtract from the hundreds part of the dividend. That is, we do 1 x 7 = __________ and then 7 – __________ = __________. We have

$$\begin{array}{r} 1 \\ 7\overline{)753} \\ \underline{7} \\ 0 \end{array}$$

– – –

7; 7; 0

196. Since we have no hundreds left over to rename as ________________, we simply "bring down" the 5 to indicate a total of __________ tens. We have

$$\begin{array}{r} 1 \\ 7\overline{)753} \\ \underline{7} \\ 05 \end{array}$$

– – –

tens; 5

197. Do we have enough tens to put at least one in each of 7 "equal" sets? __________ What is our tens partial quotient? __________ Don't forget! We must write __________ in the tens place of the numeral for the quotient so that the 1 will mean 1 __________. We have

```
   10
7 )753
   7
   05
```

No; 0; 0; hundred

198. Since the tens partial quotient is 0, we don't do the multiplication and subtraction. We simply "bring down" the 3 to see that we have __________ ones. We have

```
   10
7 )753
   7
   053
```

---

53 (or 053) Remember that 053 in this problem means the same as 53.

199. We have 53 ones. To see how many ones we can put in each of 7 "equal" sets, we look for the largest number that we can multiply by 7 and have the product less than or equal to __________.

53

200. Since 7 x 7 = __________ and 7 x 8 = __________, we see that our ones partial quotient is __________. We have

```
   107
7 )753
   7
   053
```

49; 56; 7

201. We next do the multiplication 7 (*quotient*) x 7 (*divisor*) = __________ and subtract this from __________ (*ones part of dividend*). Fill in the blanks below to complete the problem.

```
       107 ________
    7 ) 753
        7
        053
        ___
        ___
```

```
               107 R 4
    49; 53; 7 ) 753
                7
                053
                 49
                  4
```

202. To do

```
9 ) 347
```

we first look at the hundreds digit of __________ to see if we have enough hundreds to put 1 in each of __________ "equal" sets. Do we have enough? __________

347; 9; No (3 is less than 9)

203. We don't have enough hundreds so we look at the first two digits of __________. We see that we have a total of __________ tens.

347; 34

204. We look for the largest number whose product with 9 is less than or equal to __________. Since 9 x 3 = __________ and 9 x 4 = __________, our tens partial quotient is __________. Fill in the blanks below.

```
9 ) 347

   ____

   ____
```

```
                  3
34; 27; 36; 3; 9 ) 347
                   27
                    7
```

205. Complete the problem by filling in the blanks below.

```
     3 ____
9 ) 347
    27
     7 ____

    ____

    ____
```

```
     38 R 5
9 ) 347
    27
     77
     72
      5
```

Now do Exercise 5.4.

5.4 Exercise List

1)	a) $3\overline{)7}$	b) $3\overline{)73}$	c) $3\overline{)739}$
2)	a) $8\overline{)963}$	b) $4\overline{)901}$	c) $7\overline{)843}$
3)	a) $5\overline{)127}$	b) $6\overline{)124}$	c) $9\overline{)257}$
4)	a) $2\overline{)213}$	b) $8\overline{)807}$	c) $7\overline{)767}$
5)	a) $3\overline{)207}$	b) $6\overline{)495}$	c) $9\overline{)108}$

5.5 Division: Any Dividend; Divisors Less Than 10

Now that we've reviewed division of numbers less than 1000 by numbers less than 10, we will now look at problems where the dividend is any whole number and the divisor is less than 10. Recall that in the last section we compared the first steps of

```
  1          1           1
3)4        3)45        3)452
  3          3           3
  -          -           -
  1          1           1
```

The *work* of the first step of each problem was the same even though the *quantities* represented by the 4 in the numeral for the dividend are different in each problem.

Thus, we could have done the *work* without worrying about the quantities involved.

206. Consider the numeral 1234.... The three dots mean that we have left out some digits of the numeral. Thus, we (do/do not) ________________ know what number the 1 represents.

— — —

do not

207. We don't know what number the 1 of 1234... represents. However, it represents 10 of the quantity in the place held by __________ and __________ of the quantity in the place held by 3.

— — —

2; 100

208. Even though we don't know what the quantities are, the *total* is 1 of the quantity in the place held by __________, 12 of the quantity in the place held by __________, __________ of the quantity in the place held by 3, and __________ of the quantity in the place held by 4.

— — —

1; 2; 123; 1,234

209. Suppose someone said, "Start working the problem $8\overline{)1234...}$ I'll tell you the rest of the digits of the numeral for the dividend later." Impossible, you say? Let's try. We write

$$8\overline{)1234...}$$

We look at the first digit of the numeral for the dividend. We don't know what quantity is represented but we have only __________ of them.

1

210. The divisor tells us to form __________ "equal" sets. Do we have enough of the quantity represented by the first digit to put at least one of that quantity in each of 8 "equal" sets? __________

8; No

211. Now we look at the first and second digits together.

$$8\overline{)12|34...}$$

We have __________ of the quantity represented in the second place. Do we have enough of this quantity to put at least one in each of 8 "equal" sets?

____

12; Yes

212. Since 8 x 1 = __________ and 8 x 2 = __________, we can put __________ of the quantity represented in the second place in each of 8 "equal" sets. That is, the partial quotient for the second place is __________. We have

$$\begin{array}{r} 1| \\ 8\overline{)12|34...} \end{array}$$

8; 16; 1; 1

213. We next do 1 x 8 = __________ and subtract from __________ to have

```
    1¦
8 ) 12¦34...
    8¦
    -
    4¦
```

Note that we don't need to worry about what quantity is represented in the second place since we know that 1 (of that quantity) x 8 = __________ (of that quantity) and 12 (of that quantity) – 8 (of that quantity) = __________ (of that quantity).

8; 12; 8; 4

214. We have 4 "left over" of the quantity represented in the second place. We "rename" this as __________ of the quantity in the third place. We "bring down" the 3 in the third place to see that we have a total of __________ of this quantity. We have

```
    1 ¦
8 ) 123¦4...
    8 ¦
   ---
    43¦
```

40; 43

215. We think, "How many of this quantity can we put in each of 8 'equal' sets?" Since 8 x 5 = __________ and 8 x 6 = __________, our partial quotient in this place is __________. We have

```
    15¦
8 ) 123¦4...
    8 ¦
   ---
    43¦
```

40; 48; 5

216. We then do 5 x 8 = __________ and subtract from __________ to have

```
     15
8 ) 1234...
    8
    43
    40
    ----
```

40; 43; 3

217. We then "bring down" the 4 to see that we have a total of __________ of the quantity represented in that place. We have

```
     15
8 ) 1234...
    8
    43
    40
     34
```

34

218. Since 8 x 4 = __________ and 8 x 5 = __________, our partial quotient for this place is __________. Fill in the blanks below to complete this stage of the problem.

```
     154¦
8 ) 1234¦...
    8   ¦
    43  ¦
    40  ¦
     34 ¦
    ____
    ____
```

32; 40; 4

```
     154¦
8 ) 1234¦...
    8   ¦
    43  ¦
    40  ¦
     34 ¦
     32 ¦
      2 ¦
```

219. Now, we've done all the work we can without knowing the "missing digits." So we'll tell you. There was only one digit missing, and it was a 5. Complete the problem.

```
     154
8 ) 12345
    8
    43
    40
     34
     32
      2___
     ________
     ________
```

```
     1543 R 1
8 ) 12345
    8
    43
    40
     34
     32
      25
      24
       1
```

220. Now let's do 6,034 ÷ 6. We write

____) ________

---

6) 6034

221. We look at the first digit of the numeral for the dividend. This time we know the 6 means 6 ________________. However, we don't care about this. What we really care about is that whatever the quantity is, we have __________ of them.

---

thousands; 6

222. Do we have enough to put at least one in each of 6 "equal" sets? ________ What is our partial quotient for this place? ________
Fill in the blanks below.

$$\begin{array}{r} 1 \\ 6\,\overline{)\,6034} \\ \underline{} \\ \underline{} \end{array}$$

$$\text{Yes; 1; } \begin{array}{r} 1 \\ 6\,\overline{)\,6034} \\ \underline{6} \\ 0 \end{array}$$

223. We next "bring down" the ________ from the second place to see that we have a total of ________ of this quantity. We have

$$\begin{array}{r} 1 \\ 6\,\overline{)\,60\,34} \\ \underline{6} \\ 00 \end{array}$$

0; 0 (or 00)

224. Since we have a total of 0 of this quantity, our partial quotient for this place is ________. We write ________ in this place for the quotient and "bring down" the ________ from the third place of the dividend. We have

$$\begin{array}{r} 10 \\ 6\,\overline{)\,603\,4} \\ \underline{6} \\ 003 \end{array}$$

Note that we are saying "for the quotient" instead of "for the numeral for the quotient."

0; 0; 3

225. Do we have enough of this quantity to put at least one in each of 6 "equal" sets? ________________ What is our partial quotient for this place? __________ What do we write in this place for the quotient? __________ What do we do after this? ________________ We have

```
   100
6 ) 6034
    6
    0034
```

No; 0; 0; "Bring down" the 4 from the next place for the dividend.

226. Complete the problem by filling in the blanks below.

```
   100 ___
6 ) 6034
    6
    0034
    ____
    ____
```

```
   1005 R 4
6 ) 6034
    6
    0034
      30
       4
```

227. Recall that earlier in this unit we saw that in any division problem, we have dividend = (quotient x ________________) + ________________ or ________________ = (divisor x ________________) + remainder.

---

divisor; remainder; dividend; quotient

228. This gives us a method for checking our work in a division problem. We find the product of the ______ and the ______, then add the remainder. This sum must be the ______ (if all our work is correct).

--- ---

quotient; divisor (either order); dividend

229. Thus, to check $6{,}034 \div 6 = 1{,}005$ R 4, we would do

$$\begin{array}{r} 1{,}005 \\ \times \quad 6 \\ \hline \end{array}$$

then add ______ to this product. This sum should be ______.

--- ---

4; 6,034

230. Check the division $6{,}034 \div 6 = 1{,}005$ R 4. ______

--- ---

$$\begin{array}{r} 1{,}005 \\ \times \quad 6 \\ \hline 6{,}030 \end{array} \qquad \begin{array}{r} 6{,}030 \\ + \quad 4 \\ \hline 6{,}034 \end{array}$$

231. Check the division $12{,}345 \div 8 = 1{,}543$ R 1. ______

--- ---

$$\begin{array}{r} 1{,}543 \\ \times \quad 8 \\ \hline 12{,}344 \end{array} \qquad \begin{array}{r} 12{,}344 \\ + \quad 1 \\ \hline 12{,}345 \end{array}$$

232. Someone tells you that $347 \div 9 = 38$ R 2. Is he right or wrong?

______

--- ---

Wrong; $\begin{array}{r} 38 \\ \times\ 9 \\ \hline 342 \end{array} \qquad \begin{array}{r} 342 \\ +\ 2 \\ \hline 344 \end{array}$

Now do Exercise 5.5.

5.5 Exercise List

1) a) $8\,\overline{)125}$ b) $8\,\overline{)1257}$ c) $8\,\overline{)12576}$

2) a) $7\,\overline{)8530}$ b) $6\,\overline{)6134}$ c) $9\,\overline{)7695}$

3) a) $5\,\overline{)1972}$ b) $4\,\overline{)5007}$ c) $3\,\overline{)3005}$

4) Check the following divisions and write "correct" or "wrong."

a) 3,901 ÷ 7 = 557 R 2 ________________

b) 93,104 ÷ 9 = 10,354 R 3 ________________

c) 93,104 ÷ 6 = 15,517 R 3 ________________

d) 35,001 ÷ 4 = 875 R 1 ________________

e) 3,666 ÷ 5 = 733 R 1 ________________

f) 43,434 ÷ 8 = 5,429 R 2 ________________

Now do Progress Test 5.1.

5.1 Progress Test

1) How are the following divisions read?

   a) $7\overline{)28}$ __________ b) $14 \div 3$ __________

2) In

$$\begin{array}{r} 4 \text{ R } 3 \\ 8\overline{)35} \end{array}$$

   a) The divisor is ________ b) The quotient is ________

   c) The dividend is ________ d) The remainder is ________

3) What divisions are shown by the following pictures?

   a) 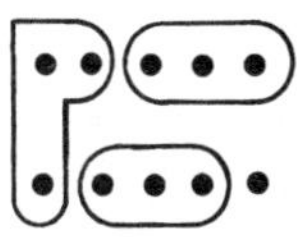 b) 

4) Fill in the other members of each "fact team."

	Multiplications		Divisions	
a)	$8 \times 7 = 56$	________	________	________
b)	________	________	$72 \div 8 = 9$	________

5) a) $16\overline{)16}$ b) $3\overline{)0}$ c) $1\overline{)17}$ d) $0\overline{)0}$

6) a) $7\overline{)47}$ b) $9\overline{)63}$ c) $8\overline{)50}$ d) $4\overline{)39}$

7) a) $5\overline{)76}$ b) $5\overline{)763}$ c) $8\overline{)297}$ d) $7\overline{)713}$

8) a) $6\overline{)7129}$ b) $9\overline{)2314}$ c) $8\overline{)1130}$ d) $3\overline{)30054}$

9) Check: $9{,}561 \div 7 = 1{,}366$ R 1. Correct or wrong?

10) a) $43 \div 6 =$ ________ b) $6 \div 43 =$ ________

5.6 Division: Multiples of Powers of 10 as Divisors

We've reviewed division of any whole number by any number less than 10. Using methods we have developed, these problems are fairly easy because we've had to do only multiplications involving basic facts (I told you that you should memorize those basic facts!). But, of course, basic facts are not the only multiplications we can do in our heads!

233. Since 6 x 9 = __________, we have 6 x 90 = __________, 60 x 9 = __________, 6 x 900 = __________, and 600 x 9 = __________.

--- ---

54; 540; 540; 5,400; 5,400

234. Since 7 x 8 = __________, we have 7 x 80 = __________, 70 x 8 = __________, 7 x 800 = __________, and 700 x 8 = __________.

--- ---

56; 560; 560; 5,600; 5,600

235. Since 4 x 5 = __________, we have 4 x 50 = __________, 40 x 5 = __________, 4 x 500 = __________, and 400 x 5 = __________.

--- ---

20; 200; 200; 2,000; 2,000

236. Now let's see how we can use the methods we've developed with some divisors greater than 10. Let's work 24,356 ÷ 90. We first write ______________.

--- ---

$90\overline{)24356}$

237. Here the divisor tells us that we are to form ______________ "equal" sets.

--- ---

90

239. Since 2 of the quantity represented in the first place is not enough to put at least one in each of __________ "equal" sets, our partial quotient for this place is __________.

90; 0

240. However, since this is the beginning of the problem, we don't write the 0. We "rename" the 2 of this quantity as __________ of the quantity in the second place. We look at the first two digits of the dividend to see that we have a total of __________ of this quantity.

$90\overline{)24|356}$

20; 24

241. Do we have enough of this quantity to put at least one in each of 90 "equal" sets? ______________ Again we don't need to indicate the __________ partial quotient for this place. We simply "rename" the 24 of this quantity as __________ of the quantity in the third place. We look at the first three digits of the dividend to see that we have a total of __________ of this quantity.

$90\overline{)243|56}$

No; 0; 240; 243

Note: When zero partial quotients happen at the beginning of a division problem, we don't have to write the 0's in the numeral for the quotient. However, as soon as we've had one non-zero partial quotient, *all later* zero partial quotients require the writing of a 0 in the numeral for the quotient.

242. We have 243 of the quantity represented in the third place of the dividend. Do we have enough of this quantity to put at least one in each of 90 "equal" sets? ______________ Thus, the first digit of the numeral for the quotient will be written above the __________ of 24356.

Yes; 3

It is often helpful to "mark" the place where the digit for the first non-zero partial quotient will be written. Of course, it's clear in the program because of the dotted lines we've been drawing for you. However, you probably won't draw the dotted lines but will simply "look" at the part of the problem you're doing (some people even "cover up" the digits of the dividend not in use). Usual ways of "marking" the place for the first digit are a small "blank," as in

$\overline{)\overset{\_}{2}4356}$, or a "check mark," as in $\overline{)2\overset{\checkmark}{4}356}$.

243. Since 90 x 2 = __________ and 90 x 3 = __________, our partial quotient for this place is __________. We have

```
       2|
90 ) 243|56
        |
```

180; 270; 2

244. We then do 2 x 90 = __________ and subtract from __________. We have

```
       2|
90 ) 243|56
        |
   _____|__
        |
   _____|__
```

```
                      2
180; 243; 90 ) 24356
               180
                63
```

245. We have 63 of the quantity in the third place "left over." We "rename" this as __________ of the quantity in the fourth place and "bring down" the 5 to see that we have a total of __________ of this quantity. We have

```
       2  |
90 ) 2435|6
     180  |
      635 |
```

630; 635

246. Since 90 x 7 = __________ and 90 x 8 = __________, our partial quotient for this place is __________. Fill in the blanks below.

```
       27
90 ) 24356
     180
      635
     ____
     ____
```

— — —

```
                    27
630; 720; 7; 90 ) 24356
                  180
                   635
                   630
                     5
```

247. We "bring down" the 6 in the fifth place of the dividend to see that we have a total of __________ of the quantity in this place. We have

```
       27
90 ) 24356
     180
      635
      630
        56
```

— — —

56

248. Do we have enough of this quantity to put at least one in each of 90 "equal" sets? ________________ Complete the problem.

```
      27 ________
90 ) 24356
     180
      635
      630
       56
```

--- ---

```
            270 R 56;   Don't forget the 0 partial quotient.
No; 90 ) 24356
         180
          635
          630
           56
```

249. We have 24,356 ÷ 90 = 270 R 56. Now check the work. __________

--- ---

```
   270        24,300
  x 90      +     56
------      ------
24,300      24,356
```

250. As we have worked division problems so far, we first look at the ________________ to see how many "equal" sets we are to form.

--- ---

divisor

251. We then look at the first digit of the ________________ to see if we have enough of that quantity to put at least one in each of the "equal" sets.

--- ---

dividend

252. In order to put at least one of a given quantity in each of the "equal" sets, we must have at least as many of that quantity as we have sets. Is any one-digit number as big as a two-digit number? ________________

— — — — — —

No

253. Thus, if the divisor is a two-digit number, we will never have enough of the quantity represented by the first digit of the ________________ to put at least one of that quantity in each of the "equal" sets.

— — — — — —

dividend

254. When our divisor is a *two*-digit number, we will look at the first *two* digits of the ________________ to see if we have enough of the quantity in the *second* place for each of the "equal" sets.

— — — — — —

dividend

255. When our divisor is a *three*-digit number, we will look at the first ________________ digits of the dividend, and, when the divisor is a *four*-digit number, we look at the first ________________ digits of the divisor.

— — — — — —

three (or 3); four (or 4)

256. To do 7,410 ÷ 30, we first write ________________.

— — — — — —

$30\overline{)7410}$

257. Since the divisor is __________ and 30 is a ________________-digit number, we look at the first ________________ digits of the dividend.

$30\overline{)74\,|\,10}$

— — — — — —

30; two (or 2); two (or 2)

258. We see that we have __________ of the quantity represented in the second place. Is this enough to put at least one in each of 30 "equal" sets? __________ What is the largest number of this quantity that we could put in each of 30 "equal" sets? __________

--- ---

74; Yes; 2 (Because 30 x 2 = 60 and 30 x 3 = 90. 74 is between 60 and 90.)

259. Fill in the blanks.

```
      2|
30 ) 74|10
       |
   ____|
       |
    ___|
```

--- ---

```
      2
30 ) 7410
     60
     14
```

260. We next "bring down" the 1 to see that we have a total of __________ of the quantity represented in the third place of the dividend. We have

```
      2 |
30 ) 741|0
     60 |
     141|
```

--- ---

141

261. Since 30 x 4 = __________ and 30 x 5 = __________, our partial quotient for this place is __________. Fill in the blanks below.

```
        24
 30 ) 741 0
      60
      141
      ____
      ____
```

```
                        24
120; 150; 4; 30 ) 7410
                    60
                    141
                    120
                     21
```

262. Complete the problem by filling in the blanks.

```
       24____
 30 ) 7410
      60
      141
      120
       21____
      __________
      __________
```

```
       247
 30 ) 7410
      60
      141
      120
       210
       210
         0
```

263. To do 24,579 ÷ 700, we first write ______________.

— — —

$700\overline{)24579}$

264. Our divisor tells us to form __________ "equal" sets. Since the divisor is a ______________-digit number, we look at the first ______________ digits of the dividend.

$700\overline{)245|79}$

— — —

700; three (or 3); three (or 3)

265. Do we have enough of the quantity represented in the third place to put at least one in each of 700 "equal" sets? ______________

— — —

No

266. Since 245 of the quantity is not enough to put at least one in each of 700 "equal" sets, we next look at the first ______________ digits of the dividend. Fill in the blanks below.

$$\begin{array}{r} \underline{\quad\;\;}| \\ 700\overline{)2457|9} \\ \underline{\quad\quad\quad} \\ \underline{\quad\quad\quad} \end{array}$$

— — —

$$\text{four (or 4); } \begin{array}{r} 3 \\ 700\overline{)24579} \\ \underline{2100} \\ 357 \end{array}$$

267. Fill in the blanks below.

```
         3____
700 ) 24579
      2100
       357____

     _________
     _________
```

```
         35 R 79
700 ) 24579
      2100
       3579
       3500
         79
```

268. Now let's work

```
400 ) 121154
```

Where will the first digit of the quotient be written? ________________

Above the fourth place of the dividend.

269. Fill in the blanks.

```
        __|
400 ) 1211|54
          |
     _____|___
          |
      ____|___
```

```
         3
400 ) 121154
      1200
        11
```

270. Fill in the blanks.

```
         3 ¦ ___
400 ) 12115¦4
      1200 ¦
        11 ¦ ___
```

```
         30
400 ) 121154
      1200
        115
```

271. Fill in the blanks.

```
         30__________
400 ) 121154
      1200
        115____

      _________
      _________
```

```
         302 R 354
400 ) 121154
      1200
        1154
         800
         354
```

Now do Exercise 5.6.

5.6 Exercise List

1) a) $30\overline{)657}$ b) $80\overline{)9765}$ c) $40\overline{)5803}$

2) a) $50\overline{)2257}$ b) $70\overline{)47603}$ c) $60\overline{)47603}$

3) a) $200\overline{)35671}$ b) $900\overline{)9765}$ c) $400\overline{)40035}$

4) a) $700\overline{)5573}$ b) $600\overline{)13140}$ c) $8000\overline{)491802}$

5) Check the divisions below.

a) 43,512 ÷ 20 = 2,175 R 12 ____________________

b) 13,501 ÷ 500 = 26 R 48 ____________________

c) 45,555 ÷ 3,000 = 15 R 555 ____________________

5.7 Any Dividend and Any Non-Zero Division

All of the division problems we have worked so far have been fairly easy because the products of partial quotients and divisors have needed only basic multiplication facts or "do-in-your-head" multiplications. However, if 42 is the divisor, (partial quotient) x 42 is not as easy.

In the next several frames we will look at some methods that will help you to divide any whole number by any other whole number.

272. Let's work the problem 3,198 ÷ 42. Here the divisor is ________.

— — — — — —

42

273. Since the divisor is 42, we know that we will have to do some multiplications of the form (partial quotient) x ________.

— — — — — —

42

274. Also, in the method of doing division that we have developed, each partial quotient is a ________________-digit number.

— — — — — —

one

275. Thus, one way of helping to do 3,198 ÷ 42 is to get a piece of "scratch paper" and work all of the multiplications of one-digit numbers x 42. Of course, we already know 0 x 42 = ________ and 1 x 42 = ________.

— — — — — —

0; 42

276.

42	42	42	42
x 2	x 3	x 4	x 5
84	____	____	210
42	42	42	42
x 6	x 7	x 8	x 9
____	294	336	____

(You may wish to copy the facts in this frame for use in the next several frames.)

126; 168; 252; 378

277. We now have all the facts we need to work 3,198 ÷ 42. We next write ________________.

---

42) 3198

278. We then observe that the divisor is a ________________-digit number. Thus, we will look at the first ________________ digits of the dividend.

42) 31 ¦ 98

---

two (or 2); two (or 2)

279. How many of the quantity represented in the second place of the dividend do we have? ________ Is this enough to put at least one of this quantity in each of 42 equal sets? ____________ What do we do next? ____________

31; No; Look at the first three digits of the dividend.

280. We look at the first three digits of the dividend.

42) 319¦8

How many of the quantity represented in the third place of the dividend do we have? __________ Is this enough to put at least one of this quantity in each of 42 "equal" sets? __________________

— — — — — —

319; Yes (Don't forget to mark where the first digit of the quotient will be written.)

281. To determine how many of the quantity represented in the third place of the dividend, we look at the multiplication we did earlier. We know that the partial quotient for this place is the largest number that we can multiply times __________ and still be less than or equal to __________.

— — — — — —

42; 319

282. From the multiplications we did earlier, 7 x 42 = __________, and 8 x 42 = __________. Thus, 8 is too large, and __________ is our partial quotient for this place. We have

7¦

42) 319¦8

— — — — — —

294; 336; 7

283. Since we've already done 7 x 42 = __________, we simply write 294 within the problem and subtract. Fill in the blanks below.

```
        7 |
  42 ) 319|8
          |
     _____|_____
     _____|_____
          |
     _____|_____
```

```
              7
294; 42 ) 3198
           294
           ---
            25
```

Be careful! We are subtracting 294 of the quantity in the third place of the dividend. Thus, the *last* digit of 294 must be in the third column.

284. We next "bring down" the 8 to see that we have a total of __________ of the quantity in the fourth place of the dividend. We have

```
        7  |
  42 ) 3198|
       294 |
       --- |
        258|
```

258

285. We go back to the earlier multiplications. Since 6 x 42 = ________ and 7 x 42 = ________, our partial quotient for this place is ________. Complete the problem.

```
      7___
42 ) 3198
     294
     ---
      258
     ______
     ______
```

```
                      76 R 6
252; 294; 6; 42 ) 3198
                   294
                   ---
                    258
                    252
                    ---
                      6
```

In the last several frames, we have showed you a method that you can use to work *any* division problem. It requires that you do eight multiplications (and know two properties of multiplication) ***but it will work***. We will now look at some methods which can cut down the number of multiplications you have to do, but if you have trouble, the method we just showed you will be available.

286. Let's do 9,976 ÷ 68. We first write ________________.

---

68) 9976

287. Since our divisor, ________, is a ______________-digit number, we look at the first ______________ digits of the dividend.

```
68 ) 99¦76
```

68; two (or 2); two (or 2)

288. How many of the quantity represented in the second place of the dividend do we have? __________ Is this enough to put at least one of this quantity in each of 68 "equal" sets? ________________ Where will the first digit of the quotient be written? ________________

--- ---

99; Yes; In the second column of the problem (or above the second 9 of 9976).

289. We must now determine our partial quotient for this place. We know that 1 x 68 = __________. We can now go to "scratch paper" and do 2 x 68.

```
 68
 x2
 ____
```

--- ---

```
68;  68
     x2
    136
```

290. Since 2 x 68 = 136 and 136 is greater than 99, our partial quotient for this place will be __________. Fill in the blanks below.

```
       1¦
68 ) 99¦76
   _____¦___
        ¦
   _____¦___
```

--- ---

```
        1
1; 68 ) 9976
        68
        31
```

Of course, most of you didn't actually need to do 2 x 68. You could look at the 99 in the first two places of the dividend and see that the partial quotient for this place is 1.

291. We next "bring down" the 7 to see that we have a total of __________ of the quantity represented in the third place. We have

```
      1 |
68 ) 997|6
     68 |
     ---
     317|
```

317

292. Now, we try to "guess" the partial quotient for this place. What would your "guess" be? __________

Of course, any answer is correct here. However, most "guesses" should be 3, 4, 5, or 6.

293. There are several ways to "guess" the partial quotient. One way is to observe that 10 x 68 = __________ and 317 is a little less than half of that.

680

294. Another way would be to think, "68 is less than 70 so that 4 68's is less than 4 70's (or __________) and 5 68's is less than 5 70's (or __________)."

280; 350

295. In either case, we can "guess" that our partial quotient will be 4 or 5. We go to "scratch paper" and do

```
68      68
x4      x5
___     ___

___     ___
```

```
68      68
x4      x5
___     ___
272     340
```

296. Since 317 is between 272 (= 4 x ________) and 340 (= ________ x 68), our partial quotient for this place is ________. Fill in the blanks.

```
        14
68 ) 997 6
     68
     317
     ______
     ______
```

- - -

68; 5; 4;

```
        14
68 ) 9976
     68
     317
     272
      45
```

297. We next "bring down" the 6 to see that we have a total of ________ ones. We have

```
        14
68 ) 9976
     68
     317
     272
      456
```

- - -

456

298. We now "guess" the next partial quotient. We can use one of the earlier methods, or we can observe that 456 is "quite a bit" greater than 340 (remember that we already did 5 x 68 = ________). Thus, 5 is too small for this partial quotient, and we would guess 6 or ________.

- - -

340; 7

299. We next use "scratch paper" to do

```
 68        68
 x6        x7
____      ____
```

408; 476

300. What is our partial quotient for this place? __________ Why? __________________ Fill in the blanks.

```
       14______
68 ) 9976
     68
     317
     272
      456
     ______
     ______
```

6; 456 is between 6 x 68 = 408 and 7 x 68 = 476. 7 is "too big."

```
       146 R 48
68 ) 9976
     68
     317
     272
      456
      408
       48
```

Note that we could have saved ourselves two multiplications with some "mental" subtraction. We might not be able to do 317 – 272 exactly in our heads, but we should be able to see that it will be less than our divisor, 68. Likewise, we should be able to see that 456 – 408 will be less than 68. This would have saved us doing 5 x 68 for the second partial quotient and 7 x 68 for the third. However, *don't expect that your first "guess" will always be right.*

301. Now, let's do $793 \overline{)243,782}$. Since our divisor is a ________-digit number, we first look at the first ________ digits of the dividend.

$793 \overline{)243|782}$

---

three (or 3); three (or 3)

302. How many of the quantity represented in the third place of the dividend do we have? ________ Is this enough to put at least one of this quantity in each of 793 "equal" sets? ________ What do we do next? ________

243; No; Look at the first four digits of the dividend.

303. We next look at the first four digits of the dividend.

$793 \overline{)2437|82}$

Since we have a total of ________ of the quantity represented in the fourth place, we have enough of that quantity to put at least one in each of ________ "equal" sets. Thus, the first digit of the numeral for the quotient will be written above the ________ of 243782.

Note: If you are working this problem on your own paper, you should mark the place where the first digit for the quotient will be written.

2437; 793; 7

304. We next "guess" the partial quotient for this place. By observing that 2,437 is "quite a bit" less than 10 x 793 = ________ or by observing that 793 is "about 800" and 2,437 is "about 2,400," we can see that 3 is a good "guess" for this partial ________.

7,930; quotient

305. We next go to "scratch paper" to do

______

x 3

______

793

x 3

2,379

306. Is 2,379 less than 2,437? ________ Doing "mental" subtraction, is 2,437 – 2,379 less than 793? ________ Is 3 the partial quotient for this place? ________ Fill in the blanks.

```
       3
793 ) 2437 82
```

Yes; Yes; Yes;

```
        3
793 ) 243782
      2379
        58
```

307. We next "bring down" the ________ to see that we have a total of ________ of the quantity represented in the fifth place of the problem. We have

```
        3
793 ) 24378 2
      2379
       588
```

8; 588

308. We have 588 of the quantity represented in the fifth place of the problem. Is this enough to put at least one of this quantity in each of 793 "equal" sets? ________________ What is our partial quotient for this place? __________ Do we have to write 0 in this place of the numeral for the quotient? ____________ ____

No; 0; Yes. We've already had a non-zero partial quotient.

309. We write 0 in the numeral for the quotient and "bring down" the __________ to see that we have a total of __________ of the quantity represented in the sixth place of the dividend. We have

```
        30   
793 ) 243782
      2379   
        5882
```

2; 5882

310. We next "guess" the partial quotient for this place. What would your "guess" be? __________

Your "guess" should be 7 or 8. Even with a smaller divisor, 700, 9 would be too great a partial quotient (because 9 x 700 = 6,300), and with a larger divisor, 800, 6 would be too small a partial quotient.

311. We go to "scratch paper" to do

```
 793
x  7
____
```

```
 793
x  7
5,551
```

312. Without actually subtracting, we can see that 5,882 – 5,551 will be less than ________, and we will not have to do 8 x 793. Our partial quotient for this place is ________. Complete the problem.

```
        30_____
793 ) 243782
      2379
        5882
       ______
       ______
```

```
                307 R 331
793; 7; 793 ) 243782
               2379
                 5882
                 5551
                  331
```

We should note at this point that many people do their multiplications "within the problem" instead of using "scratch paper." Thus, in doing the problem we just worked, they "guess" the partial quotient and write

```
        3
793 ) 243782
```

They then do 3 x 793, writing the digits of the product as they do the multiplication to get

```
        3
793 ) 243782
      2379
```

This is fine if you are able to do the multiplication this way. It will save some time. However, if your "guess" partial quotient is wrong, you have to erase and start over.

313. To do 213) 12141, we first look at the first ________ digits of the dividend. We want to see if we have enough of the quantity represented in the third place of the dividend to put at least one in each of ________ "equal" sets.

---

three (or 3); 213

314. How many of the quantity in the third place of the dividend do we have? __________ Do we have enough to put at least one of this quantity in each of 213 "equal" sets? ________________

--- ---

121; No

315. We look at the first four digits of the ________________.

213) 1214¦1

We have a total of __________ of the quantity represented in the fourth place of the dividend. Since this is enough to put at least one in each of __________ "equal" sets, the first digit of the numeral for the quotient will be written above the ________________ of 12141.

--- ---

dividend; 1214; 213; 4

316. Using one of the methods we have for "guessing" partial quotients, many people "guess" 6 as partial quotient for this place. Do the multiplication 6 x 213.

____

--- ---

```
  213
x   6
-----
1,278
```

317. We have 6 x 213 = __________, and 1,278 is (greater/less) ________________ than 1,214. Thus, our "guess" for partial quotient was (too large/too small/correct) ________________. What should our partial quotient for this place be?

____

--- ---

1,278; greater; too large; 5

318. We go to "scratch paper" to do

```
213
x 5
-----
```

--- ---

1,065

319. Fill in the blanks.

```
         5
213 ) 1214 1
      ______
      ______
```

```
         5
213 ) 12141
      1065
       149
```

320. Complete the problem by filling in the blanks below.

```
         5___
213 ) 12141
      1065
       149____
      ______
      ______
```

```
         57
213 ) 12141
      1065
       1491
     - 1491
          0
```

We have pointed out several ways to "guess" partial quotients. Most people use some form of rounding of the divisor to help with their "guesses." The more you practice, the better your "guesses" should become. However, if you are willing to do the eight multiplications (2 x divisor, 3 x divisor, etc.), you can do division problems without "guessing."

Now do Exercise 5.7.

5.7 Exercise List

1) a) $\begin{array}{r} 68 \\ \times 2 \\ \hline \end{array}$ b) $\begin{array}{r} 68 \\ \times 3 \\ \hline \end{array}$ c) $\begin{array}{r} 68 \\ \times 4 \\ \hline \end{array}$ d) $\begin{array}{r} 68 \\ \times 5 \\ \hline \end{array}$

   e) $\begin{array}{r} 68 \\ \times 6 \\ \hline \end{array}$ f) $\begin{array}{r} 68 \\ \times 7 \\ \hline \end{array}$ g) $\begin{array}{r} 68 \\ \times 8 \\ \hline \end{array}$ h) $\begin{array}{r} 68 \\ \times 9 \\ \hline \end{array}$

2) Now use problem 1 to help you do the following divisions.

   a) $68\overline{)752}$ b) $68\overline{)3907}$ c) $68\overline{)69017}$

   Use any method to do the following.

3) a) $21\overline{)799}$ b) $53\overline{)4986}$ c) $72\overline{)43848}$ d) $212\overline{)7423}$

4) a) $47\overline{)1695}$ b) $98\overline{)8036}$ c) $39\overline{)20677}$ d) $398\overline{)17913}$

5) a) $65\overline{)1776}$ b) $86\overline{)8430}$ c) $54\overline{)20574}$ d) $138\overline{)7728}$

6) If 27 children share 15,066 sunflower seeds equally, each will get ___________ seeds.

5.8 Applications

Now that we've reviewed division of whole numbers, we will look at ways of using division in "everyday" problems. We will use both of our ways of thinking about division and also the relationship between multiplication and division.

321. Seven people ate at a restaurant. The total cost was $37.66. If each person was to pay the same amount, how much should each pay? Here we are to start with __________ and form __________ sets of "equal" size.

— — — — — —

$37.66; 7

322. That is, we are to do the division problem $37.66 ÷ __________.

— — — — — —

7

323. Just as with addition, subtraction, and multiplication, we work division problems involving dollars and cents as if we were using whole numbers, then place the dollar sign and point properly. Thus, to do $37.66 ÷ 7, we first write

$$\underline{\quad\quad}\overline{)\,37.66}$$

Note: We usually don't write the symbol "$" inside the division sign.

— — — — — —

7

324. Now, there are two ways to proceed. One is to proceed as if the point isn't there. Do the division

$$7\,\overline{)\,37.66}$$

as if the point isn't there.

— — — — — —

$$\begin{array}{r} 5\,38 \\ 7\,\overline{)\,37.66} \\ \underline{35} \\ 26 \\ \underline{21} \\ 56 \\ \underline{56} \\ 0 \end{array}$$

325. When we use this method, we then write a "$" in front of the numeral for the quotient and mark a point between the hundreds and tens places in the numeral for the quotient to have

```
   $ 5.38
7 ) 37.66
    35
     26
     21
      56
      56
       0
```

Thus, $37.66 ÷ 7 = __________.

— — —

$5.38

326. Now, let's do $42.75÷9. We first write ______________.

— — —

```
9 ) 42.75
```

327. This time we look at the second way of proceeding. We place the point in the numeral for the quotient before doing the division. As we can observe from the last problem, the point in the quotient is placed directly above the point in the dividend. We have

```
       .
9 ) 42.75
```

We now do the division as before and write the "$" in front of the numeral for the quotient. Complete the problem.

— — —

```
   $ 4.75
9 ) 42.75
    36
     67
     63
      45
      45
       0
```

328. A farmer's chickens laid 408 eggs one day. The farmer must ship the eggs in cartons of one dozen (12) each. How many cartons does he use? Here we are to start with __________ eggs and form "sets of __________."

--- ---

408; 12 (or 1 dozen)

329. Here we do the division 408 ÷ __________.

--- ---

12

330. How many cartons did the farmer use? ______________

--- ---

$$\begin{array}{r} 34 \text{ cartons} \\ 12\overline{)408} \\ \underline{36} \\ 48 \\ \underline{48} \\ 0 \end{array}$$

331. A construction company removed 6,831 cubic feet of dirt in building a new supermarket. To find the number of cubic yards they moved, we must first remember that there are 27 cubic feet in one cubic yard. We then ask how many groups of __________ cubic feet we can make out of 6,831 cubic feet.

--- ---

27

332. We can get the answer to the question, "How many groups of 27 cubic feet can we get from 6,831 cubic feet?" by dividing __________ cubic feet by __________ cubic feet.

--- ---

6,831; 27

333. Find the number of cubic yards in 6,831 cubic feet. ________________

— — —

```
      253
27 ) 6831
     54
     143
     135
      81
      81
```

There are 253 cubic yards in 6,831 cubic feet.

334. The *average* of 5 numbers is found by dividing the sum of these numbers by 5. The average of 6 numbers would be found by dividing the sum of these numbers by __________.

— — —

6

335. The average of 12 numbers would be found by dividing the ________________ of these numbers by __________.

— — —

sum; 12

336. The average of 3 numbers is found by ________________.

— — —

dividing the sum of these numbers by 3

337. If Joe is 62 in. tall, Pete 65 in. tall, Mike 58 in. tall, and Dave 63 in. tall, then the average height of these 4 boys is found by dividing the ________________ of these numbers by 4.

— — —

sum

338. Find the average height of the 4 boys given above. ______________

```
  62          62
  65      4 ) 248
  58          24
 +63          08
248 in.        8
               0
```

62 in. is the average height of these 4 boys.

339. A boy had scores of 87, 93, 78, 82, and 75 on 5 math tests. What was his average score? __________

```
 87        83
 93    5 ) 415
 78        40
 82         15
+75         15
415          0
```

340. We know that $9 \times 8 =$ __________. Thus, $72 \div 8 =$ __________, and $72 \div 9 =$ __________. Similarly, since $D = r \times t$, where D is *distance* traveled, r is the average *rate* of speed and t is the *time* traveled, we also have $r = D \div t$ and $t = D \div$ __________.

72; 9; 8; *r*

341. If a man drove 576 miles in 12 hours, what was his average rate of speed?

____

$r = D \div t = 576 \div 12$

```
     48
12 ) 576
     48
      96
      96
       0
```

His average rate of speed was 48 miles per hour.

342. Four college students drove to Florida during their vacation. Taking turns driving, they traveled 966 miles at an average rate of speed of 42 miles per hour. How long did the trip take? ________________

— — —

$t = D \div r = 966 \div 42$

$$\begin{array}{r} 23 \\ 42\,\overline{)\,966} \\ \underline{84} \\ 126 \\ \underline{126} \\ 0 \end{array}$$

The trip took 23 hours.

343. We have seen that the area, A, of a rectangle is given by $A = l \times w$, where l is the *length* of the rectangle and w is the *width* of the rectangle. Thus, we can show that $l = A \div$ __________ and $w =$ __________ $\div\, l$.

— — —

w; A

344. The area of a rectangle is 13,764 square inches. The width of the rectangle is 74 inches. What is the length of the rectangle? ________________

— — —

$l = A \div w = 13{,}764 \div 74$

$$\begin{array}{r} 186 \\ 74\,\overline{)\,13764} \\ \underline{74} \\ 636 \\ \underline{592} \\ 444 \\ \underline{444} \\ 0 \end{array}$$

The length of the rectangle is 186 inches.

Now do Exercise 5.8.

5.8 Exercise List

1) A certain giant ape in his latest film squashed 26 cars (small imported cars) at every stomp. If the movie studio had only 234 cars for the whole movie, how many times could he stomp? ____________

2) The studio in problem one had a budget of $247.25 for hiring extras to play a terrified mob fleeing the city. If they use 43 people, how much will each get for the scene? ________

3) A prehistoric monster that comes out of the sea wrecked 17 boats in his first film, 29 boats in the sequel, 77 boats in his next movie (it was a holiday weekend), and five boats in his last film (which had a very low budget). What was the average number of boats wrecked per movie?____________

4) In a recent flop movie, "Almost Around the World in 76 Days," the characters traveled 24,700 miles in 76 days. What was their average rate of speed?

____

5) The red carpet at the airport is 3 feet wide. If its area is 78,813 sq. ft., how long is it? ____________

6) On the way back the characters in problem 4 took a short cut and only traveled 24,624 miles. They took a plane with an average rate of 342 miles per hour. How many hours did they travel? ____________ How many days?

____

Now do Progress Test 5.2.

5.2 Progress Test

1) a) $70\overline{)856}$ b) $50\overline{)4350}$ c) $60\overline{)6239}$

2) a) $400\overline{)8971}$ b) $900\overline{)38700}$ c) $3000\overline{)70053}$

3) a) $42\overline{)863}$ b) $79\overline{)8941}$ c) $56\overline{)4371}$

4) a) $68\overline{)3581}$ b) $95\overline{)1800}$ c) $102\overline{)73112}$

5) a) 1,578 ÷ 156 = __________ b) 3,000 ÷ 15 = __________

6) Check the following and write "correct" or "wrong."

a) 3,941 ÷ 33 = 119 R 14 b) 3,050 ÷ 231 = 13 R 37

7) A man who went out to eat 5 days in a row spent $13.95 for a steak dinner on Sunday, $1.56 at a hamburger joint, $1.35 at a cheaper place, $.79 at a taco stand, and $.25 for an orange drink on Thursday (since he was broke). What was his average per day? __________

8) In an impressive new building, one rectangular office has 1,904 square feet of carpet. If the width of the carpet is 34 feet, how long is it? __________

9) The halls in that building are so long that the secretaries use motorized roller skates. If one of them makes a 12,320 yard trip in 7 minutes, what is her average rate? __________

10) If there are 696 secretaries to be evenly divided among 174 offices, there are __________ secretaries per office.

You have now completed Unit V of *Basic Mathematics*. If you are to take a test over this unit, you should ask for it now.

5.1 Exercise List—Answers

1)

Two Ways of Writing Division		How They Are Read
21 ÷ 4	4) 21	21 divided by 4
13 ÷ 5	5) 13	13 divided by 5
31 ÷ 7	7) 31	31 divided by 7
29 ÷ 3	3) 29	29 divided by 3
37 ÷ 7	7) 37	37 divided by 7

2) a) 7
   b) 37
   c) quotient
   d) remainder

3) a) dividend
   b) divisor
   c) 2
   d) 3

4) a) 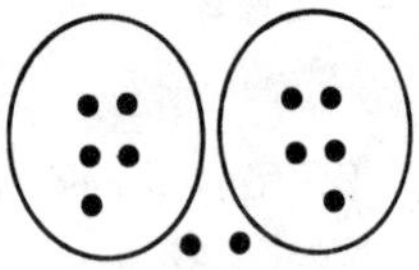

12 ÷ 5 = 2 R 2

b) 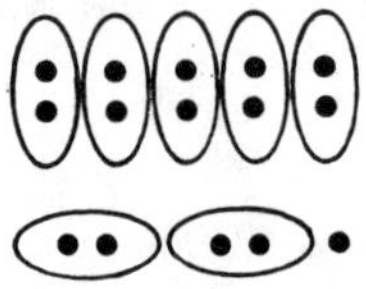

15 ÷ 2 = 7 R 1

c) 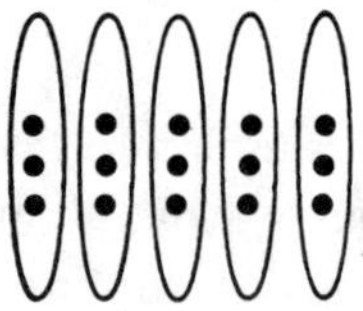

15 ÷ 3 = 5

d) 

$$\begin{array}{r} 3 \text{ R } 4 \\ 7\overline{)25} \end{array}$$

5.2 Exercise List—Answers

1)

Multiplication facts		Division facts	
3 x 7 = 21	7 x 3 = 21	21 ÷ 3 = 7	21 ÷ 7 = 3
1 x 14 = 14	14 x 1 = 14	14 ÷ 1 = 14	14 ÷ 14 = 1
3 x 9 = 27	9 x 3 = 27	27 ÷ 3 = 9	27 ÷ 9 = 3
4 x 9 = 36	9 x 4 = 36	36 ÷ 4 = 9	36 ÷ 9 = 4
12 x 3 = 36	3 x 12 = 36	36 ÷ 12 = 3	36 ÷ 3 = 12

2) a) 1 b) 13
c) 0 d) undefined

3) a) 15 b) 0
c) undefined d) 1

4) Multiplication fact — Division fact

a) 4 x 5 = 20 or 5 x 4 = 20 — 20 ÷ 4 = 5 or 20 ÷ 5 = 4

b) 4 x 4 = 16 — 16 ÷ 4 = 4

c) 2 x 8 = 16 or 8 x 2 = 16 — 16 ÷ 2 = 8 or 16 ÷ 8 = 2

5) a) 27 ÷ 3 = 9 seats b) 27 ÷ 9 = 3 trainloads

5.3 Exercise List—Answers

1) a) 4 R 2 b) 6 R 3 c) 5 R 2

2) a) $\begin{array}{r} 4\text{ R }1 \\ 4\overline{)17} \\ \underline{16} \\ 1 \end{array}$ b) $\begin{array}{r} 4\text{ R }3 \\ 8\overline{)35} \\ \underline{32} \\ 3 \end{array}$ c) $\begin{array}{r} 5\text{ R }5 \\ 9\overline{)50} \\ \underline{45} \\ 5 \end{array}$ d) $\begin{array}{r} 9\text{ R }2 \\ 3\overline{)29} \\ \underline{27} \\ 2 \end{array}$

3) a) 6 R 2 b) 7 R 4 c) 8 R 6 d) 9 R 3

4) a) $4\overline{)6}$ = 1 R 2; 4, remainder 2

b) $6\overline{)4}$ = 0 R 4; 0, remainder 4

c) $3\overline{)8}$ = 2 R 2; 6, remainder 2

d) $8\overline{)3}$ = 0 R 3; 0, remainder 3

It looks as if division does not have the commutative property, since 6 ÷ 4 is not equal to 4 ÷ 6!

5) a) 27 ÷ (9 ÷ 3)

= 27 ÷ 3

= 9

b) (27 ÷ 9) ÷ 3

= 3 ÷ 3

= 1

So division doesn't have the associative property either!

6) a) $9\overline{)56}$ = 6 R 2 56 = (6 x 9) + 2

b) $4\overline{)3}$ = 0 R 3 3 = (0 x 4) + 3

c) $3\overline{)22}$ = 7 R 1 22 = (7 x 3) + 1

5.4 Exercise List—Answers

1) $3\overline{)7}$ = 2 R 1; 6, remainder 1

b) $3\overline{)73}$ = 24 R 1; 6, 13, 12, remainder 1

c) $3\overline{)739}$ = 246 R 1; 6, 13, 12, 19, 18, remainder 1

```
2) a)   120 R 3        b)   225 R 1        c)   120 R 3
      8 ) 963             4 ) 901             7 ) 843
          8                   8                   7
          --                  --                  --
          16                  10                  14
          16                   8                  14
          --                  --                  --
           03                 21                   03
                              20
                              --
                               1
```

```
3) a)   25 R 2         b) 20 R 4        4) a)   106 R 1       b) 100 R 7
      5 ) 127          c) 28 R 5              2 ) 213         c) 109 R 4
          10                                      2
          --                                      --
           27                                      13
           25                                      12
           --                                      --
            2                                       1
```

5) a) 69 b) 82 R 3 c) 12

5.5 Exercise List—Answers

```
1) a)   15 R 5         b)   157 R 1        c) 1,572
      8 ) 125             8 ) 1257
          8                   8
          --                  --
          45                  45
          40                  40
          --                  --
           5                   57
                               56
                               --
                                1
```

2) a) 1,218 R 4
   b) 1,022 R 2
   c) 855

3) a) 394 R 2
   b) 1,251 R 3
   c) 1,001 R 2

4) a) (7 x 557) + 2
3,899 + 2
3,901
correct

b) (9 x 10,354) + 3
93,186 + 3
93,189 not 93,104
wrong

c)
```
  15,517     93,102
x      6   +      3
--------   --------
  93,102     93,105
wrong
```

d)
```
  875      3,500
x   4    +     1
-----    -------
3,500      3,501
wrong
```

e)
```
  733      3,665
x   5    +     1
-----    -------
3,665      3,666
correct
```

f)
```
 5,429     43,432
x    8   +      2
------   --------
43,432     43,434
correct
```

5.6 Exercise List—Answers

1) a)
```
      21 R 27
30 ) 657
     60
     --
      57
      30
      --
      27
```

b)
```
     122 R 5
80 ) 9765
     80
     --
     176
     160
     ---
      165
      160
      ---
        5
```

c)
```
     145 R 3
40 ) 5803
     40
     --
     180
     160
     ---
      203
      200
      ---
        3
```

2) a)
```
      45 R 7
50 ) 2257
     200
     ---
      257
      250
      ---
        7
```

b) 680 R 3
c) 793 R 23

3) a)
```
       178 R 71
200 ) 35671
      200
      ---
      1567
      1400
      ----
       1671
       1600
       ----
         71
```

b) 10 R 765
c) 100 R 35

4) a) 7 R 673 b) 21 R 540 c) 61 R 3,802

5) a)
```
  2,175     43,500
x    20   +     12
-------   --------
 43,500     43,512
correct
```

b)
```
     26     13,000
x   500   +     48
-------   --------
 13,000     13,048
wrong
```

c)
```
     15     45,000
x 3,000   +    555
-------   --------
 45,000     45,555
correct
```

5.7 Exercise List—Answers

1) a) 136 b) 204 c) 272 d) 340
   e) 408 f) 476 g) 544 h) 612

2) a)

```
      11 R 4
68 ) 752
     68
     --
      72
      68
      --
       4
```

b)

```
       57 R 31
68 ) 3907
     340
     ---
      507
      476
      ---
       31
```

c)

```
      1014 R 65
68 ) 69017
     68
     --
      101
       68
      ---
       337
       272
       ---
        65
```

3) a) 38 R 1
   b) 94 R 4
   c) 609
   d) 35 R 3

4) a) 36 R 3
   b) 82
   c) 530 R 7
   d) 45 R 3

5) a) 27 R 21
   b) 98 R 2
   c) 381
   d) 56

6)

```
        558 seeds for each
27 ) 15066
     135
     ---
      156
      135
      ---
       216
       216
       ---
```

5.8 Exercise List—Answers

1)

```
        9 stomps
26 ) 234
     234
     ---
```

2)

```
       $ 5.75
43 ) 247.25
     215
     ---
      322
      301
      ---
       215
       215
       ---
```

3)

```
 17
 29
 77
  5
---
128
```

```
   32 boats
4 ) 128
    12
    --
    08
     8
    --
```

4)

```
     325 miles per day
76 ) 24700
     228
     ---
      190
      152
      ---
       380
       380
       ---
```

5)

```
   26271 feet
3 ) 78813
    6
    -
    18
    18
    --
     08
      6
     --
      21
      21
      --
       03
        3
       --
```

6)

```
      72 hours
342 ) 24624
      2394
      ----
       684
       684
       ---
```

```
    3 days
24 ) 72
     72
     --
```

5.1 Progress Test—Answers

1) a) 28 divided by 7
   b) 14 divided by 3

2) a) 8
   b) 4
   c) 35
   d) 3

3) a) 10 ÷ 3 = 3 R 1
   b) 15 ÷ 5 = 3
   or 15 ÷ 3 = 5

4) a) 8 x 7 = 56   7 x 8 = 56   56 ÷ 7 = 8   56 ÷ 8 = 7
   b) 8 x 9 = 72   9 x 8 = 72   72 ÷ 8 = 9   72 ÷ 9 = 8

5) a) 1
   b) 0
   c) 17
   d) undefined

6) a) 6 R 5
   b) 7
   c) 6 R 2
   d) 9 R 3

7) a) 15 R 1
   b) 152 R 3
   c) 37 R 1
   d) 101 R 6

8) a) 1,188 R 1
   b) 257 R 1
   c) 141 R 2
   d) 10,018

9) 1,366 × 7 = 9,562

9,562 + 1 = 9,563, not 9,561, so it is wrong.

10) a) 7 R 1
b) 0 R 6

5.2 Progress Test—Answers

1) a) 12 R 16
b) 87
c) 103 R 59

2) a) 22 R 171
b) 43
c) 23 R 1,053

3) a) 20 R 23
b) 113 R 14
c) 78 R 3

4) a) 52 R 45
b) 18 R 90
c) 716 R 80

5) a) 10 R 18
b) 200

6) a) 119 x 33 = 3,927
3,927 + 14 = 3,941
correct

b) 13 x 231 = 3,003
3,003 + 37 = 3,040
wrong

7) $13.95
1.56
1.35
.79
.25
$17.90

5) 17.90 = $ 3.58
15
29
25
40
40

8) 34) 1904 = 56 feet
170
204
204

9) 7) 12320 = 1760 yards per minute
7
53
49
42
42

10) 174) 696 = 4 secretaries
696